Spacefaring

SPACEFARING

The Human Dimension

ALBERT A. HARRISON

UNIVERSITY OF CALIFORNIA PRESS

Berkeley / Los Angeles / London

University of California Press
Berkeley and Los Angeles, California

University of California Press, Ltd.
London, England

Library of Congress Cataloging-in-Publication Data

Harrison, Albert A.
    Spacefaring : the human dimension / Albert A. Harrison.
        p.   cm.
    Includes bibliographical references and index.
    ISBN 0-520-22453-1 (cloth : alk. paper).
    1. Manned space flight.   2. Astronautics—Human factors.
    3. Space colonies.   4. Interstellar travel.   I. Title.

    TL1500 .H37 2001
    629.45—dc21   00-061522

Manufactured in the United States of America
10 09 08 07 06 05 04 03 02 01
10 9 8 7 6 5 4 3 2 1

The paper used in this publication meets the minimum requirements
of ANSI/NISO Z39.48-1992(R 1997) (*Permanence of Paper*).

For Andy Adams,
soldier, scientist, physician

Contents

Extravehicular Activities * Role Loading * THE SPACEFARER'S TOOL
KIT * Work Spaces * Basic Tools * PARTNERING WITH INTELLIGENT
MACHINES * Assigning Tasks to People and Machines * Trust * Who's
in Charge Here? * CONCLUSION *

Preface

Spacefaring is a partnership involving technology and people. This book looks at the human side of the partnership: why people are willing to brave danger and hardship to establish a human presence in space, how human behavior and culture have shaped past and present missions, and how they may shape future missions as well. Our journey begins with the earliest flights and looks forward to space tourism, space settlement, and interstellar travel, but the emphasis is on the next steps toward human occupation of space: the completion of the International Space Station, a return to the Moon, and the arrival of humans on Mars. In addition to taking a close look at spacefarers themselves—how they are selected, how they are trained, how they live and work, what they do on furlough and after retirement—I will also consider the broader organizational and political contexts that shape human progress toward the stars.

The book begins with the question of human motivation: why should people be interested in space in the first place? Space advocates point to the scientific and educational advantages of human space exploration. We are beckoned by an endless stream of scientific projects, including the prospects of conducting astronomy on the Moon and looking for signs of life on Mars. Space exploration is a focal point for engaging children in science and developing the intellectual and human resources so crucial for the success of the next generation. Space offers economic opportunity, through beaming inexpensive power to Earth, mining the asteroids, and otherwise taking advantage of the cheap materials and unusual environmental conditions that exist in space. Perhaps outer space will accommodate human expansion and ameliorate the problems that flow from overtaxing Earth's resource base. Beyond this, many people believe that space exploration will help us grow psychologically and spiritually, perhaps offering us endless renewal on a never-ending frontier.

Chapter 2 develops a framework for understanding the psychological and social dimensions of spacefaring. Human factors, narrowly defined, are human capabilities and limitations in relation to jobs, machines, and work environments. Human factors engineers design equipment, arrange workspaces and tools, and establish procedures that promote efficient performance. Human factors broadly defined extend to include personality, interests, attitudes, social relations, and culture. Human factors broadly defined comprise the focal point of this book. To understand the future of humans in space, I draw on studies of people in space itself, autobiographical and biographical material on spacefarers, and documentary accounts. Recent missions, especially the Mir and shuttle-Mir missions, have added tremendously to our knowledge base. We have also learned from human experience in spaceflight-analogous environments such as submarines, polar outposts, and other settings characterized by isolation, confinement, and other conditions that we associate with space.

Outer space is lethal and improvident. Transportation to and from outer space is extremely dangerous, and we should rejoice that so few spacefarers have lost their lives thus far. These harsh conditions shape spacecraft designs, life support systems, equipment and supplies, and regimens for preserving life and health. Because we have not evolved in Earth orbit, on the Moon and on Mars we must be inventive just to survive.

Chapter 3 explores some of the basic hazards of spaceflight, including acceleration, microgravity, and radiation. Among the undesirable biomedical consequences of life in space are space adaptation syndrome, muscular (including cardiovascular) deconditioning, altered immune systems, bone demineralization, and radiation poisoning. Certainly these effects are daunting, but so far there have been no "showstoppers," because we understand the courses of these conditions and have adequate-to-good countermeasures to prevent them from spiraling out of control, at least during missions that currently are on the drawing board. The chapter considers how to equip spacefarers to deal with everyday health problems (such as toothache and appendicitis), as well as with accidental injuries and the medical conditions (such as space adaptation syndrome) that we associate with space itself.

Chapters 4 and 5 extend the discussion of keeping people alive and well in space. Chapter 4 begins with a retrospective and prospective look at spacecraft and habitats, including the U.S. and Russian spaceships of the Mercury through Apollo eras; and space shuttles and space stations, including Skylab, Salyut, Mir, and the International Space Station. These examples help us understand requirements for safe and secure transit vehicles and habitats that can withstand temperature extremes, vibration, corrosive dust, and other threats. This introduction helps us understand the evolution of life support systems that maintain a satisfactory atmosphere, keep temperature and humidity within tolerable limits, guarantee adequate supplies of water and food, and assure acceptable levels of hygiene. Over time, humans will become less reliant on expensive resources imported from Earth and more reliant on inexpensive resources already available in space.

Habitability refers to the quality of life within an environment. Habitability depends to a greater degree on the specifications of the spacecraft and to a lesser degree on the attitudes and expectations of its occupants. Initial voyagers to previously uninhabited regions are forced to accept minimalist conditions, but those who come later require better accommodations. Chapter 5 reviews volumetric requirements, environmental legibility, windows and viewports, privacy, aesthetics, decor, and other factors that determine the quality of life in space. For years to come, it will be more difficult to assure a high quality of life in space than on Earth, but in the distant future space may be the more congenial setting.

Chapter 6 explores the technical, emotional, and social strengths required to do well in space. Initially the province of white male test pilots, space is now home to men and women from many different backgrounds and with many different interests. Today, selection teams seek candidates who are in good health and physically fit but not necessarily fitness buffs, who are exceptionally competent in their technical fields, who are highly motivated and emotionally stable, and who can get along with one another. Once chosen, candidates are trained in the operations of the spacecraft as well as in how to conduct scientific experiments or do other assigned tasks. Psychologists understand the conditions that promote effective learning, and we will see how these are incorporated into U.S. and Russian training programs. The goal is more than merely to assemble a collection of highly trained individuals: it is to build a team that performs flawlessly as an integrated whole.

Danger, deprivation, isolation, confinement, and other elements of life in space sometimes yield adverse effects. These include impaired problem-solving ability, declining vigilance, altered time perception, increased immersion in fantasy, heightened suggestibility, sagging motivation, social withdrawal, and depression. In Chapter 7 we will see that these consequences are not invariable and that isolation and confinement also have some positive effects. Like everyone else, spacefarers have many lines of defense against stress. Crewmembers can and often do help one another alleviate stress, and ground-based psychological support groups have helped both astronauts and cosmonauts. Severe psychiatric problems could erupt in space, but there are ways to keep these problems under control.

"Alone together," spacefarers are cut off from the normal network of family, friends, and social obligations and are forced to get along with colleagues who cannot be escaped for more than brief periods of time. Chapter 8 explores group composition, with special attention to the effects of mixing men and women of different ages and from different backgrounds and cultures. I will trace the implications for spaceflight of traditional and contemporary models of leadership. This discussion suggests that crew leaders must possess interpersonal as well as technical skills, and know when and how to exert authority. Social tensions can run high under conditions of isolation and confinement, and this has implications for relations with external parties (such as mission control), as well as implications for relations among the spacefarers themselves.

Tasks that are relatively easy to perform on Earth become difficult in space. Chapter 9 describes how work is complicated by microgravity, the close confines of the habitat, bulky life-support gear, and mental states. Sometimes spacefarers are exhausted by schedules that are fine on Earth but are too fast paced for space. Other times, spacefarers' attention drifts because they have too little to do or are assigned mind-numbing, repetitive tasks. Spacefaring has always been a high-tech operation, but it has become increasingly so in recent years. Understanding work in space requires an in-depth look at the partnership between people and machines. At some point spacefarers may be assisted by androids and by "tool rooms" that use nanotechnology to manufacture almost anything they need from scratch.

Given the high risks of spaceflight, both U.S. and Russian programs have a remarkable safety record during the first half century of spaceflight. Nonetheless, human lives have been lost, and unless we remain earthbound we may expect additional fatalities in the future. Chapter 10 describes how psychological factors (such as imperfect training, fatigue, and lapses of attention), small-group factors (such as communication and coordination), and organizational factors (such as red tape and pressures to "get on with it") contribute to accidents. There are, however, a number of strategies that either reduce the likelihood of accidents or curtail their adverse effects, and as we review a disheartening litany of things that could go wrong, we must always remember that more often than not people recover from equipment failures and from their own mistakes.

Even spacefarers need balance in life. Chapter 11 discusses self-maintenance activities (such as eating, sleeping, and sex) and a broad spectrum of recreational activities ranging from simple downtime to attempts at self-improvement. We consider also spacefarers' relationships with their families and friends, their activities between missions, and life after retirement.

The next three chapters look to our possible future in space and are necessarily somewhat speculative. Nonetheless, we can see how psychological and cultural factors have shaped our vision of the future and forecast, in very general ways, how people might adjust to flights and missions that have very different requirements than do the missions of today.

Chapter 12 explores how tourism, one of the world's largest industries, might expand into space. Space tourists may follow the same general path as professional spacefarers: suborbital flights, orbital flights, and interplanetary voyages. Chapter 13 reviews selected proposals to establish human communities on the Moon, on Mars, and in orbit around Earth. Conditions will be difficult at first, but according to space advocates, cheap, plentiful energy and abundant raw materials coupled with high technology will allow us to create prosperous new societies there. We shall see how, as presently envisioned, these communities would be models of social engineering as well as triumphs of technology. Chapter 14 examines procedures to assure that interstellar voyagers depart for promising destinations and provides thumbnail sketches of propulsion systems that may be capable of covering interstellar distances within acceptable periods of time. I will compare slowships, which are large comfortable spacecraft that meander lazily toward their destinations, and fastships, the equivalent of nuclear-propelled Winnebagoes that could complete interstellar journeys in relatively few generations. The chapter toys with the possibility that breakthrough physics might allow us to cover interstellar distances in the course of a single lifetime and takes a prospective look at interstellar humanity: how human life and civilization might evolve if our descendants spread throughout the galaxy.

Judging on the basis of projections made decades ago, humans should have returned to the Moon by now and would be about to set forth for Mars, if they had not already arrived. The concluding chapter looks at the belief that the space program has stalled. In actuality, over the past forty years people have done much to develop the economic and technical infrastructure required for long-term space habitation, and have gained immense experience in space itself. Nonetheless, a broad array of sociopolitical, organizational, economic, and psychological factors make it difficult for humanity to take the next steps. To accelerate progress, we must develop efficient launch technologies and reduce waste. Accelerated progress will require developing new partnerships and increasing the attractiveness of investment in space.

NASA is a complex, multifaceted organization with many achievements. Certainly, the Apollo Moon landings at the end of the 1960s represent one of the greatest technological and human triumphs of all time, but one part of this complexity is a certain level of ambivalence toward the social psychology of space exploration. On the one hand, NASA has convened panels, encouraged research in spaceflight-analogous environments, sponsored conferences and literature reviews, and in other ways encouraged thought on psychological and social adaptation to space. On the other hand, for reasons I shall explore in chapter 2, astronauts are discouraged from speaking out about the problems they encounter in space. Largely for purposes of public relations, NASA has been highly protective of the astronauts' sparkling image and has discouraged studies that could help humanity prepare for the long-term habitation of space. The penalty for this reluctance to explore the human side of spaceflight is evident in the memoirs of some astronauts and in many of the events that occurred when U.S. astronauts served on the Russian space station Mir.

We owe the spacefarers a better understanding of the human side of spaceflight. We owe this to spacefarers past because only through recognizing the true human cost do we acknowledge the full enormity of their contributions. We owe this to spacefarers present to help guarantee their safety, ease their work, and assure them high overall quality of life. And we owe this to spacefarers future, who might be a little less capable, a little less tough, and a little less resolute than their predecessors, and who might openly welcome anything that we do to ease their transition to space.

Acknowledgments

In a sense, this book originated in 1978, when Mary M. Connors of NASA Ames Research Center invited me to help her identify human requirements for extended spaceflights. Over the next two decades, my thinking on this topic was shaped especially by Dr. Connors and our colleague Faren Akins but by other collaborators as well, including Chris McKay and Yvonne A. Clearwater, also of NASA Ames, and successive generations of UC Davis students, including Dan Bout, Barrett Caldwell, Steve Franzoi, Kathy Hoyt, Nancy Struthers, Joshua Summit, and James Moulton Thomas. Many of the ideas contained in this book stem from these collaborations and from informal discussions with campus colleagues and with behavioral scientists and many other people fascinated by human factors in space. These include Karen and Poul Anderson, Greg Bennett, Robert T. Bigelow, Marilyn Dudley-Rowley, Ben Finney, Martyn Fogg, Jim Funaro, D. M. Harland, Phil Harris, Nick Kanas, Larry Lemke, Mark Lupisella, John Carter McKnight, Tom Meyer, Jim Miller, Edgar D. Mitchell, Gerald Nordley, Alcestis Oberg, Larry Palinkas, Doug Raybeck, Reed Reiner, John Spencer, Jack Stuster, Harvey Wichman, Richard Zimmer, the NASA Aerospace Education Specialists, and many others, who will find some of their ideas reflected in this work.

Several people provided generous assistance by commenting on part or all of the manuscript. I am indebted to Jim Lowe, Declan O'Donnell, Larry Penwell, John P. Schuessler, Don Scott, and James Moulton Thomas for sharing their broad range of expertise and providing detailed comments on selected chapters, in some cases several chapters. Thanks also to Eric W. Davis, James Oberg, Peter Suedfeld, Allen Tough, and several anonymous reviewers who provided thoughtful commentary on entire drafts.

Howard Boyer, my editor at University of California Press, deserves high grades for seeing strengths in a preliminary version of the manuscript, for facilitating the formal review process, and for cutting me enough slack to do this project in the way that I saw fit. In addition, I am indebted to Danielle Jatlow for providing strong staff support. Jean McAneny, the production editor, achieved a delicate balance between no nonsense and graciousness, and Bonita Hurd has my admiration for finding countless ways to improve a manuscript that I had thought was already perfect. Most of all I am grateful to my partner in life, Mary Ann Harrison, who not only put up with me while I poured immense amount of time into this project but provided line-by-line commentary on various drafts and proofread as well.

This work draws on many books and articles, which I have tried to accurately represent and fully acknowledge. Despite voluminous background literature and the generous assistance of so many people, and despite strenuous efforts to maintain the highest level of vigilance, a book of this length is bound to contain some errors and shortcomings. Any deficiencies in this book are mine and I freely acknowledge this.

CHAPTER 1

WHY SPACE?

For several months during 1997, the world riveted its attention on Russia's Mir Space Station. Successor to a string of Salyut stations, Mir had been launched eleven years before. Arguably the world's first true space station (the United States' Skylab had not been intended for continuing occupancy), Mir offered previously unparalleled challenges and opportunities for humans in space. Over the years a succession of Russian cosmonauts had gone about their business, conducting science, trying new commercial applications, and setting records for time aloft. Beginning in 1995 the cosmonauts were joined in turn by the U.S. astronauts Norman E. Thagard, Shannon W. Lucid, John E. Blaha, Jerry Linenger, C. Michael Foale, David Wolf, and Andrew Thomas in a dramatic show of East-West cooperation and in anticipation of the International Space Station (ISS) soon to come.

Mir performed its mission well, but in 1997, after more than a decade in space, Mir struck some observers as cranky, something like an aging car.[1] The problems started in late February, when, as a cosmonaut tried to activate a chemical canister, a fire broke out. In March, an oxygen generator failed and a seven-ton resupply ship was unable to dock. In April the cooling system developed a leak, forcing the shutdown of an air filtration system and causing nasal congestion among the crew. Problems escalated in June when another cargo ship collided with Mir and punched a hole in the Spektr module where metallurgical research was done. Also that month a power problem caused the batteries to run low, and the station's computer disconnected from the control system. In July there was another air leak, and, more ominously, the stabilizing gyroscopes that kept the station oriented toward the sun shut down, making it impossible for the solar panels to absorb the necessary energy. By midmonth Commander Vasily Tsibliyev's heartbeats were irregular when he was under the stress of exercise. Was his high level of stress a cause or a result of the escalating problems?

Two days later Mir started to drift off course after the accidental disconnection of a computer cable and another power shortage. August was hardly better, as it was marked by failing oxygen generators, a malfunctioning automatic pilot system, and yet another main computer breakdown. In September the main computer failed twice, once on September 22, just days before the space shuttle *Atlantis* was scheduled to dock in order to retrieve Michael Foale, whose tour of duty had concluded. It seemed as if everything that could go wrong, did.

Working in space is always difficult due to cramped quarters, temperature extremes, and the problems associated with weightlessness. Certainly, during the summer of 1997 the pressures on the spacefarers were magnified by the malfunctions and by certain knowledge that they were under close scrutiny by the masses who followed their progress on television and radio. Yet despite equipment failure and human error, the spacefarers' training and determination prevailed. As had always been the case on Mir, the problems were corrected. No lives were lost and the Mir crew carried on.

Mir and its crew were not the only ones at risk during these difficult months. The cascading problems threatened United States–Russian collaboration and perhaps even the near-term future of human spaceflight. No Mir, no ISS; no ISS, no trip to Mars. Anxious space enthusiasts watched as NASA's inspector general evaluated the situation to decide whether or not another astronaut should join Mir. Astronauts accepted their missions and served with distinction. Only a little over half a year after Andrew Thomas departed from Mir, the first components of the ISS were delivered to orbit.

Early 1997 was not the first time that human intelligence, flexibility, and motivation prevailed over incipient catastrophes in space. Just under thirty years earlier, the crew of Apollo 13 managed to circle the Moon and return safely to Earth after the explosion of an oxygen tank left them with barely enough electric power, air, and water to survive. During Mir's widely publicized problems, some spectators castigated the crew and gloated over possible recriminations following their return to Earth. These critics missed an essential point: to err is human, but to recover is human too. On Mir, as on Apollo 13 a generation earlier, when all was said and done the best part of human nature prevailed. As William K. Douglas, the physician to America's earliest astronauts, once said, too often we point to examples of human frailty when we should see the more prevalent signs of greatness instead.[2]

Space exploration is an intrinsically human activity. An automated satellite or robot probe, a contemporary flight aboard a shuttle or Mir, the developing ISS, a return to the Moon, and our first footsteps on Mars all rest on human motives and require human abilities and skills. Many of us, when we contemplate space exploration, think of huge rockets belching clouds of smoke and fire, satellite-tracking dishes, complex communications systems, and of course, the spacecraft themselves. Yet the incredible advances of the last century that first made it possible for heavier-than-air flight, and then to put men and women into space in almost routine fashion, represent far more than a triumph of technology. These accomplishments reflect human ingenuity, adaptability, and determination and are harbingers of greater achievements to come.

THE BECKONING HEAVENS

The stars have always called us, but only for the past forty years or so have we been able to respond. First, people went one by one, and then in groups of two, three, or more. First, space was the province of white male test pilots, but today space draws men and women with many different backgrounds, from many different lands. First, people went for hours, then days, and now for weeks and months. Some day we will go there to stay.

Space has been the province of the selected few: as of the year 2000, only about four hundred people had flown there. Yet, for each person who visits space, many more stand ready. Thousands respond to each call for astronauts, and for every one who applies to become a spacefarer, there must be scores who dream about visiting space.

At present, space travel is extremely expensive. According to one recent estimate, it costs approximately ten thousand dollars to put one pound in low Earth orbit using the space shuttle, and about four thousand dollars to put a pound in orbit using conventional rockets.[3] For the spacefarers themselves, the risks and personal costs are high. People who want to become spacefarers must pass stiff competitions and undergo extensive training. They may have to master a difficult foreign language and culture before they can participate in an international mission. It may be years, if ever, before they are assigned a flight. In the course of their careers, between training, flight, and public relations tours, they are rarely home with their families.

By normal terrestrial standards, life in space is extremely dangerous. To leave Earth's gravity, spacefarers ride atop tons of burning materials, and they perhaps undertake difficult docking maneuvers when reaching their destinations. Typically, today's spacefarers live in noisy, cramped conditions and forego most of the amenities that are regularly available on Earth. There, they maintain difficult and relentless work schedules, perhaps for months at a time.

Why is it, then, that so many people are willing to meet the challenge? This is particularly intriguing in that the next generation of spacefarers, like previous generations, will consist of bright, educated people who would be assured a secure, comfortable, and prosperous existence on Earth. And why are societies sometimes willing to devote enormous amounts of resources to spaceflight (during the 1960s the United States devoted up to 5 percent of its annual budget to spaceflight)?[4] Space advocates argue that we go to space to learn, to tap resources and develop wealth, and to grow and prosper as individuals and as a species.

KNOWLEDGE MOTIVES

We are an inquisitive species. We have the time and intellectual resources to generate and disseminate knowledge. Since antiquity, our ancestors have wondered about the heavens, and over the past few centuries we have developed the tools to help satisfy our curiosity. Space exploration teaches us about the universe. Over the years, space programs have sponsored far-ranging theoretical and applied research, and they have given us wonderful tools for engaging people's interests in science.[5]

Advancing Science and Technology

Science and technology gain from the basic research that is a precondition for both robotic and crewed missions. Our movement into space has been accompanied by advances in the physical sciences and engineering. Our desire to send humans into space has forced us to improve our understanding of biology and medicine and to develop life support systems for air and water recycling, temperature and humidity control, food production and storage, and waste management. Spacecraft and satellites provide wonderful platforms for observing and learning about Earth and for unraveling the mysteries of the universe. Unlike their terrestrial counterparts, whose efficiency is undermined by atmospheric distortion, orbiting telescopes are remarkably effective for their size. Space telescopes such as the Hubbell permit observations that would otherwise be impossible. We learn also from robot probes that give us close views of neighboring planets, that sometimes land and analyze local conditions and even return samples to Earth. As R. C. Parkinson points out, space exploration allows us to address such big philosophical questions as "What is the origin of Earth and the solar system?" "What is the origin of the universe?" and "What is the role of consciousness in the universe?"[6]

Although people tend to focus on the adventurous aspects of the Apollo voyages to the Moon, Paul D. Lowman Jr. and David M. Harland add that the voyages brought us excellent scientific returns. These expeditions were complex scientific affairs that involved remote sensing, geologic mapping, and placement of monitoring instruments that lasted for years, as well as collection of 384 kilograms of Moon rocks, which are still undergoing analysis.[7] As a result of Apollo we know much more about the origin and nature of the Moon: despite its rough and unfriendly appearance, it could be a habitable and useful world.

Space offers us conditions that are valuable for certain kinds of experimental research. These include extreme cold, high vacuum, immense uncluttered areas, and a degree of remoteness that could insulate humanity from an experiment gone wrong. The biggest drawing card is microgravity (also known as o-G, or weightlessness), which is useful for research in metallurgy, crystallography, and chemistry, including pharmaceuticals.

Education and Human Resource Development

Space exploration fuels people's interest in science, technology, and nature. Space and space-related activities grab children's attention and are wonderful tools for education. Bruce Cordell and Joan Miller recommend developing space education programs to reinforce students' interest in space, help them separate fact from fiction, and encourage them to think analytically.[8] They suggest beginning with children in the third grade, when they are able to begin grasping the necessary concepts. A good program requires continuing efforts on the part of the teachers, coupled with presentations by expert guests. Presentations should be relatively brief—twenty minutes for the younger children, one hour for high school students. Speakers can engage interest by stressing danger and the unknown, exploration and discovery, and far-away places. Slides or other visual materials are useful, and good humor is essential. Students' experience with science fiction such as *Star Trek* is a good point of departure for separating fact and fiction.

Strenuous educational efforts are undertaken by the not-for-profit Challenger Centers, established by K-12 teachers nationwide as a memorial to Sharon Christa McAuliffe, who, following a brief moment as the first teacher on a space flight, died in the 1986 *Challenger* explosion. Working in partnership with schools, universities, museums, and other institutions, Challenger Centers "use the theme of space exploration to create positive learning experiences, foster interest in science, math, and technology, and motivate young people to explore."[9]

Center staff give teachers on-line resources and conduct educator workshops, as well as work directly with children. Some regional centers have shuttle or other mock-ups that allow students to take part in simulated missions. Volunteers themselves assemble annually to hear guest speakers, engage in workshops, and increase their own command of the material.

NASA has always maintained education and information programs. These include programs that disseminate information to the press and the public, and workshops for teachers. Each year groups of teachers gather at NASA centers for a two-week program on science and science education.

NASA brings science education to the schools via the Aerospace Education Specialist Program contracted through Oklahoma State University. This involves a squad of thirty-six specialists, most of whom are assigned to states or urban communities and all of whom are supported by an enthusiastic and productive staff. These specialists work with state leaders in education as well as with teachers and students. You may encounter them on the road driving large white vans from school to school. Their otherwise nondescript vehicles are stuffed with items such as thin slivers of Moon rocks embedded in clear plastic disks, global positioning units, and imitation space suits. There are cartons containing various displays and brochures and the many personal effects required to sustain the vagabond teachers. These educators are skilled at engaging the interest of primary and middle school students, and they encourage give-and-take as students learn about nature. In a slow year, they visit one thousand teachers and twenty thousand students.

Programs such as these offer two benefits. First, they sensitize students to the importance and value of space exploration. Second, they encourage students to become trained in science and technology. Thus, as educational efforts help prime the next generation of citizens to support human activities in space, they help prepare the workforce necessary to bring these activities about.[10]

ECONOMIC MOTIVES

Centuries ago, people imagined fabulous wealth beyond the seas. Today, we envision fabulous wealth beyond our skies. Telescopes, spectrometers, interplanetary probes, and other tools confirm these resources' presence. Science and education are important in our culture, but we will require significant economic returns if we wish to justify the extremely high cost of establishing a growing and continuing human presence in space.

Some of the riches from activity in space are already in hand, and some should be attainable in the near future. Others, especially those that depend on crewed spaceflight, are beyond our reach and may remain so indefinitely. As we anticipate harvesting cheap electric power, mining valuable minerals, and establishing luxury resorts for tourists and similar ventures, we may overlook the fact that accessing these riches will be extremely difficult and expensive. In a sense, we are like a child with a tiny allowance daydreaming about expensive mountain bicycles in a store window. Under such conditions it can be very difficult to conduct an honest cost-benefit analysis or develop a realistic time line. Overpowered by the grandeur of the opportunities that glitter before us, we may lose sight of the fact that it may be quite some time before we are able to seize them.

All aspects of space exploration—whether it be constructing or operating telecommunications satellites, conducting cutting-edge astronomy with the Hubble space telescope, or establishing a strip-mining operation on the Moon—have immediate economic benefits. So far, not one dollar has been spent in space—all money spent on space exploration has been spent right here on Earth. According to some analyses, every dollar spent on the Apollo Moon Program translated into seven to eight dollars returned to the economy in new goods and services.[11] Space-related activities create high-level jobs: for scientists, engineers, and technicians, for analysts and accountants—for the people who will fly in space and the people whose work on Earth supports them. Scott Sacknoff and Leonard David estimate that parts of the space industry are growing at rates surpassing 20 percent annually, thus creating forty thousand new jobs each year.[12]

Spin-Offs

Space exploration has encouraged the development of new technologies that have translated into industrial and consumer products that enrich our lives on Earth. These are the so-called spin-offs of the space program. According to Paul S. Hardesen, by the mid-1990s NASA claimed over thirty thousand of them.[13]

Some of the better-known products include Velcro; thin, lightweight blankets with amazing insulation properties; and a ballpoint pen that writes upside down, on grease, and irrespective of atmospheric pressure. Other spin-offs are "invisible" in that their origin is not widely known. For example, the requirement for onboard computers for navigation and automatic piloting moved us away from mainframe computers and helped make pocket calculators and personal computers available to us. The clunky but reliable onboard computers so essential to Apollo were the forerunners of the minicomputers that control automobile engines and serve as the nervous system for hundreds of products, including "smart" toys. Some already existing products such as Teflon and Tang (a powdered orange drink) became famous due to their association with the space program.

Lightweight, transportable medical packs developed in space are useful in other hard-to-reach locations, such as Antarctica. Other medical spin-offs include implantable medication systems and sensors; automatic defibrillators; intensive care telemetry; computer-enhanced angiography; and synthetic, portable speech prosthetics.[14] Research intended to help spacefarers grow crops with minimal amounts of water could help us conserve water on Earth, and studies of waste management in space could help us clean up some of the world's greatest cities.[15]

Managing Life on Planet Earth

Our activities in space help us take care of planet Earth. Salient here are satellite surveillance systems that allow people to monitor weather and other environmental changes. These generate knowledge useful for managing agricultural production as well as addressing environmental disasters.

Satellites play an essential role in navigation and communication. Formerly difficult problems of navigation are turned into child's play. On navy ships the compass and the sextant are supplemented, if not replaced, by global positioning units, or GPUs, with their liquid crystal displays. GPUs base their readings by locating themselves relative to the known positions of a carefully orchestrated fleet of satellites in 12-hour circular orbits. If the GPU gets good radio reception from a minimum of three of these satellites, it can compute its precise geographic location and elevation. As with most electronic devices, the price has plummeted over the years, and now they are affordable by small-boat owners, motorists, and hikers, as well as owners of aircraft carriers. The most technologically advanced GPUs are so accurate that for some civil engineering purposes the GPU has replaced both the theodolite and the laser as the preferred tool for surveying.

Satellites enable near-instantaneous communication around the world. They are essential for news and entertainment, whether the program is directed through a commercial broadcasting station or beamed directly to individual consumers. They are a crucial link in person-to-person communication, allowing people to talk over vast distances at low cost. In the foreseeable future everyone who owns a cellular-type phone may be able to communicate with everyone else directly through a satellite. This means, for example, that a person in the Yukon may be able to call someone driving through the Sahara Desert.

Launching satellites for the $10–$12 billion satellite communications industry is relatively inexpensive and borders on the routine. This is a useful lesson here for those of us who consider crewed spaceflight too expensive and difficult to become profitable. The original Sputnik was slightly larger than a basketball and smaller than many communications satellites. Yet, when Russia launched Sputnik more than forty years ago, it took the world by storm. Each early U.S. attempt to launch a small satellite was front-page news; now, unless you subscribe to a specialty publication you most likely won't even hear about a communications satellite launch. Sending people into space will never be as cheap or easy as launching a small communications satellite, but the point remains: what seems challenging if not impossible right now may seem not all that difficult later on.

Use of Space Resources

Enthusiasts point out that space houses abundant resources that could help us overcome shortages here on Earth. Although the oil scares of the 1970s are behind us, Earth's population continues to grow at an alarming rate and our energy consumption keeps apace. According to an analysis by William H. Siegfried, between the beginning of our current century and 2050, the United States' demand for electrical energy will increase from 3 to 9 trillion kilowatt-hours, an increase by a factor of three![16] Even today, many less developed countries either cannot support their electrical power needs or do so by burning precious and highly polluting fossil fuels. The more developed countries have access to nuclear power, but it may be reaching a plateau. Further utilization of nuclear power is limited by its potentially hazardous nature, resistance on the part of environmental activists, and the risk that nuclear materials will fall into the hands of terrorists.

Sanders D. Rosenberg and John S. Lewis, among others, describe how we may be able to use the Moon or satellites to collect solar energy in space and then beam it back to Earth in the form of microwaves.[17] In one scenario, the solar power would be collected on the Moon's surface and then beamed to Earth. In another, lunar resources would be used to construct a large number of satellites in geosynchronous Earth orbits. Power would be beamed from these satellites to Earth. An advantage of this strategy is that we could eliminate many of our largest power-distribution grids, recycling the metals and improving the appearance of our land. Instead of, say, running power lines from Niagara Falls to Boston, we could beam the microwaves to Boston, where the power is consumed.

Another technique for meeting Earth's future power needs begins with mining helium-3 on the Moon. This substance, which is rare on Earth, is abundant on the Moon. After it is mined and brought to Earth it can be fused with deuterium in a process that involves little radioactivity and waste. The Moon has enough helium-3 to accommodate today's energy requirements for another thousand years. When, several hundred years from now, the Moon's supply begins to run out, we may be able to obtain raw materials from the gas giants—Jupiter, Saturn,

Uranus, and Neptune. In effect, these planets have enough raw materials to meet our energy needs forever.

Lewis describes also the incredible mineral wealth embodied in asteroids.[18] By his estimate, there is enough iron in the asteroid belt to meet Earth's needs for 400 million years. The small asteroid Amun, with its two-kilometer diameter, would yield about $8,000 billion in iron and nickel; $6,000 billion in cobalt; and another $6,000 billion in platinum and related metals such as osmium, iridium, and palladium. Ignoring many other precious ingredients, Amun's current value on Earth would be about $20,000 billion. Of course, we cannot reach Amun right now, and in any case the sudden influx of a huge quantity of rare metals would cause their value to plummet. Still, it is nice to know that these resources may be available to our descendants.

Some asteroids, Lewis claims, are more accessible to us than the Moon. These include the most dangerous asteroids, those that are rapidly approaching and could collide with Earth. Rather than trying to destroy them with explosives, we should mine them. In the course of stripping them down, we would gain new resources as we eliminate the threat. Gregory Matloff points out that space station technology would allow us to visit asteroids that approach Earth. Because this would not require descending into and climbing out of another planet's gravity well, it might be easier for us to visit a near-Earth asteroid than to visit Mars.[19]

A very long time from now we may use space resources to support ourselves, if we develop new societies on huge orbiting platforms, on the Moon, or on Mars. Large-scale emigration into space could offer us two benefits. First, by dispersing ourselves throughout the solar system our species can survive meteor strikes or other major calamities that could eradicate life on Earth. If we disperse widely enough, we can even survive the eventual death of our Sun. Second, emigration will allow human population to grow beyond the size that can be accommodated on Earth.[20] Our solar system can accommodate perhaps 10 quadrillion people. If and when we once again start rubbing elbows with one another, we may be able to migrate to other stars.

Space Tourism

Tourism is one of the world's largest industries, and at some point we may expect to send tourists into space. Several companies already hope to vie for tourists' dollars, and some are even accepting reservations for suborbital flights. Later on, tourists may orbit Earth two or three times, spend a couple of days on an orbiting hotel, or even circle the Moon. Right now we have no civilian spaceplanes, no orbiting hotels, no beckoning oases on the Moon or Mars. Still, surveys suggest that people want to visit space, recognize that it will be expensive, and are willing to save for their trip.[21]

It might seem that space tourism would be one of the last industries to develop, long after we have established orbiting factories and strip-mining on the Moon. After all, tourists will not be attracted by the primitive conditions that scientists and explorers accept. However, space tourism advocates such as David Ashford and Patrick Collins believe that tourism may be one of the *first* space industries to emerge and that it will then pave the way for everything else.[22] The key to all human endeavors in space is developing low-cost methods for getting there. The lure of big financial returns from the huge tourist market gives entrepreneurs incentive to do this. Ashford and Collins themselves developed phased plans for first bringing very wealthy tourists into space, then expanding the opportunity to a broader range of people, and finally bringing the cost down to approximately ten thousand dollars per person for orbital trips. In the course of developing ways to mass-market space vacations, they hope to bring down the costs for everyone—engineer, scientist, construction worker—and open the door for industrialization and settlement.

PSYCHOLOGICAL AND SOCIAL MOTIVES

The great philosophers and scientists who envisioned our movement into space considered it a natural, evolutionary step, a way to become better off. Many of the German scientists who worked on the V-2 military rocket used to bombard London in World War II had their sights set on the stars. Hermann Oberth foresaw a two-pronged approach to fulfilling our destiny: outward migration of humans into space and eventual contact with intelligent beings from other worlds.[23]

Marsha Freeman stresses the work of Krafft A. Ehricke, a German rocket pioneer who arrived in the United States along with Wernher von Braun shortly after World War II.[24] His theory of astronautics, whose origins lie in the Renaissance, posits that our greatest limits are those that we place upon ourselves, that we have every right to make use of as much of the universe as we possibly can, and that as we move outward to the Moon and beyond we will fulfill our destiny.[25] He argued that, if we "think small" technologically, we consign ourselves to a closed system: one whose resources will continually erode and eventually be exhausted, leaving us in a state of entropy. If, on the other hand, we embrace technology and establish ourselves in space, we will have endless opportunity for growth. This type of thinking has developed into a "school" that propounds expanding into space as a means for overcoming the limits to human growth.[26]

The anthropologist Ben R. Finney and astrophysicist Eric M. Jones observe that humans are exploring animals.[27] Unlike many other animals born with an interest in exploring their environments, we maintain this curiosity well into adulthood. The "first giant leap for mankind," they write, "came over five million years ago, when our tiny hominid ancestors left the sheltering trees of the African forests and moved into the grasslands." As our forebears spread into more and more ecological niches, their survival depended not only on evolution but also on culture: the ability to organize socially and use tools to get work done. The emerging capacity to gather food helped our ancestors survive in the grasslands, and later their ability to form hunting bands allowed them to migrate north.

Our predecessors spread out over forests, grasslands, and jungles, moved from one continent to another, and became seafarers. By A.D. 1000, the Polynesians had settled most islands in the Pacific and the Norse were spreading into North America. Today, we inhabit polar regions and deserts and everything in between, and stand on the threshold of space. Finney and Jones note that we should think of settling space not as fantasy, technology gone wild, or imperialism, but as something basic to our nature. "We are," they note, "the exploring animal who, having spread over our natal planet, now seeks to settle other worlds."[28]

Drawing on the work of the anthropologist Joseph Campbell, Larry Lemke sees fascination with space as a reformulation of a pervasive myth, that of the hero's journey.[29] In various forms this myth is found at many times in many cultures. It is evident in the stories of Jason and the Argonauts, King Arthur, and the gold prospectors of 1849; in the Apollo program; and in the movie Star Wars. This myth is important because it helps us understand the nature of the world, ourselves, and our relationships to one another. Today, heroes venturing into space represent this basic myth reformulated for our time and our culture. Space exploration supports a belief system that incorporates ideas about the cosmos, modern applied science and technology, and the emergence of a global community. The myth of the journeying hero is, according to Lemke, crucial for our "growing up as a species."

In *Living and Working in Space,* Philip R. Harris uses the term *space ethos* in reference to cultural attitudes toward space.[30] The nature and strength of a country's space ethos depends upon many things, including the attitudes of the political leadership, the level of space activism and popular support, financial stakes, and the like. A strong space ethos is much more evident in cultures that have space programs (such as Canada, China, Japan, Russia, the United States, and the European Space Agency nations) than in countries that do not. Russia had, and still has, a strong space ethos, but following dissolution of the Soviet Union the spiritual underpinnings became less evident and economic stakes became more evident. In the United States, the very strong space ethos of the 1960s weakened after the Apollo astronauts had reached the Moon.

Although many countries do not include spacefaring among their priorities, and despite differences in space ethos among nations, Harris sees evidence of an emerging world space ethos. This was stimulated by the International Space Year of 1992 that promoted space activities throughout the world: conspicuous collaboration between the United States, Russia, and other spacefaring nations; news and entertainment that drew space activities to everyone's attention; and many other forces. Harris argues that, "driven by destiny to extend human civilization beyond our own Solar System, the bold journey through space satisfies our spirit to know, while increasing our coping skills and expanding human culture."[31]

Personal Motivation

At some point in the future, working in space may be more lucrative than a terrestrial job, but this is not true right now. The highly selected pilots, engineers, physicians, and scientists in the astronaut corps would have been able to draw salaries at high-tech companies, major hospitals, and airlines much higher than the $48,000 to $103,000 that they were paid in the late 1990s; these sums, in turn, would seem princely to the Russian cosmonauts.[32] Some people will get very rich from space, but these are more likely to be the investors than the people who actually do the work there. Social recognition is another powerful motivator. There is still considerable glamour in spacefaring. Yet the accolades accorded today's spacefarers fall far short of those accorded the astronauts and cosmonauts of yesteryear. For example, you might be able to remember the name of the first astronaut on the Moon, but how many of the other Moon-walkers can you recall? The earliest spacefarers were big-time celebrities, but most of today's astronauts and cosmonauts are not as widely known.

Perhaps it is the inherent or intrinsic satisfaction of spacefaring that is the biggest draw. Training for and working in space allows people to develop their abilities, gain a sense of accomplishment, grow psychologically, and feel worthwhile. Discussing his experiences in space, Walter Cunningham stated, "It has caused me to confidently seek a challenge wherever I can find one, to charge ahead and never look back. . . . that feeling of omnipotence is worth all that it takes to get there."[33]

Such positive motivational forces are evident in Frank White's survey of about two dozen astronauts and cosmonauts.[34] Despite acknowledging occasional inconveniences and difficulties, astronauts emphasized the significance and importance of their work and outlined profound positive effects upon their lives. White coined the term *overview effects,* which, like psychologist Abraham Maslow's "peak experiences," include a sense of transcendence, oneness with nature, and universal brotherhood.[35] Thus, looking at Earth from the Moon, Eugene Cernan reported that what he viewed was "almost too beautiful to grasp. There was too much logic, too much purpose—it was too beautiful to have happened by accident."[36] The astronaut Ed Gibson said, "You can see how diminutive your life and concerns are compared to other things in the universe. . . . it gives you inner peace."[37] In a recent discussion of his five months aboard Mir, the U.S. astronaut Jerry Linenger tells how entering orbit marked his exhilarating transition from Earthling to spaceman.[38] His months in space broadened his perspective, taught him that people are 99.9 percent alike, and made conflicts on Earth seem senseless. He learned that we can overcome setbacks and obstacles and emerge better for it. Overall, his experiences changed his life profoundly, and memories of the launch, the risks, and the accomplishments will remain with him on his deathbed.

As described in his book *The Way of the Explorer,* Edgar D. Mitchell also underwent a major transformation in space.[39] Mitchell was born in Texas and brought up on a farm near Roswell, New Mexico, partly during the depression. He lived with people who were necessarily concerned with practical matters and not afraid of working with their hands, and he acquired many of their values. After completing college, he joined the navy, in which he served with distinction as a combat fighter pilot. Later, he completed a doctoral degree in physical science at the Massachusetts Institute of Technology.

Mitchell flew on Apollo 14 and was part of the two-person lunar-landing party. During this trip, especially the return to Earth, he gained a profound sense of "universal connectedness." He sensed that the cosmos and everyone in it were all part of a deliberate plan and wondered if the cosmos itself was in some way conscious. It was a "grand epiphany" accompanied by exhilaration, and something not expected of a former Texas farmer! As a result, he became more introspective and devoted his life to reconciling inner experience and the world of science. After leaving NASA, he formed the Institute of Noetic Science to study issues of experience, science, and religion. Mitchell adds that other astronauts—not necessarily known for eloquence and humanistic interests—became involved in art, poetry, and religion after their flights.

The Apollo astronaut Harrison Schmitt, who advocates developing the Moon, stresses the importance of "being there": of direct, firsthand experience. "Whether it is reaching the crest of a mountain, attaining the top of Mount Everest, standing on the rim of the Grand Canyon, or orbiting Earth and looking down, being there," he writes, "provides a uniquely important and memorable experience."[40] In Schmitt's view, such experiences are not only personally rewarding, they are important for the evolution of humankind.

Uniting Humanity

Certain triumphs in space have brought to all of us a strong sense of human achievement.[41] The pull of Apollo, notes Harrison Schmitt, was the opportunity to watch people strive to succeed, to surpass themselves, to accomplish.[42] Worldwide, more than 600 million people sat glued to their television sets watching the first lunar landing, and the event itself induced a brief sense of world unity. As Michael Collins reported, after Apollo 11 the reaction not only of Americans but of people around the world was, "We did it."[43]

Space exploration began as a form of competition between the two superpowers, the United States and the Soviet Union, then slowly evolved into a collaborative international venture. Over the years, and in cooperation with the U.S. Department of Space, NASA has negotiated hundreds of collaborative agreements with scores of foreign countries. Canada invested $100 million in developing the remote manipulator arm for the shuttle, and the European Space Agency (ESA) invested $1 billion in Spacelab, a laboratory facility that fits within the shuttle's bay.[44] In addition to the United States and Russia, fourteen other nations are taking an active part in developing the International Space Station. There is also a long and rich tradition of international collaboration in automated missions, such as the Cassini mission to Saturn, launched in 1997.

International cooperation serves many purposes. It cuts costs by eliminating the duplication of effort that would occur if, for example, two or three competing nations each built its own spaceport. It reduces the financial burden placed upon any individual nation by spreading the total cost across many nations. A balky president or the U.S. Congress might be willing to move forward with a new space program if other nations were willing to pick up part of the tab. International cooperation allows nations that could not afford space exploration on their own to enter the spacefaring arena. The involvement of many different nations can lend stability to a program: although a nation might be tempted to quit, it is under moral and diplomatic pressure not to do so.[45] However, this is not always enough to ensure that all parties follow through on their commitments.

Perhaps the most important reason for collaborating in space is to reduce international tensions. A common justification for a joint United States–Russian mission throughout the history of the space program was to underscore cooperation rather than competition between the two countries and reduce East-West tensions.[46] As senator and then future U.S. president Lyndon Baines Johnson remarked to the UN General Assembly in 1958, "Men who have worked together to reach the stars are not likely to descend together into the depths of war and desolation."[47]

CONCLUSION

We seek to expand our presence in space in order to conduct science and pursue knowledge, tap the fabulous resources of the universe, grow psychologically, and help resolve international tensions. Some of these dividends for human involvement in space are available right now, some may be available in the foreseeable future, some might not materialize for a very long time, if ever. Collectively these potential payoffs have gained the attention of explorers, adventurers, settlers, technologists, merchants, and profiteers, among others.[48]

Larry Lemke divides motives for space exploration into two categories, utilitarian and transformational.[49] Utilitarian motives include science and industry, national defense, generation of wealth, and other practical activities that have preoccupied us in the past. Although space enthusiasts may view these activities as bigger or better than past activities of a similar type, they are qualitatively similar to what we have already done and tend to perpetuate the status quo. Transformational motives include developing human potential, eliminating conflict and war, and promoting social renewal by offering people a fresh start. These represent the next step forward, evolving beyond creatures of the world to become creatures of the universe. Transformational motives are qualitatively different from business as usual: they involve innovation and will change the status quo.

In their attempts to gain support from other people, space advocates tend to stress utilitarian goals. They emphasize projects that will please practical, everyday people and that will help keep the politicians and "bean counters" under control. But the reality, states Lemke, is that whatever the promise of tomorrow, today's forays into space do not yield a sufficient number of tangible payoffs to justify the extraordinary costs. Certainly, communications and other automated satellites constitute a thriving business, but space power systems, manufacturing, asteroid mining, and other commercial activities have yet to get off the drawing board. For many people it is the transformational motives that keep the dream alive.

Perhaps we go to space because we are hardwired to identify new and unusual situations and to make sense of them. The Sun, the Moon, and what lies beyond are part of our environment, and as residents of the solar system and Milky Way some of us would like to have a better idea of what is available outside of Earth's gravity well. As a species, we humans have always wondered about the planets and stars, but only recently have we been able to venture off our planet's surface. This biological foundation is strengthened by several cultural themes, including our interest in technology, our fascination with exploration and conquest, and our hope for new opportunities.

CHAPTER 2

SPACEFLIGHT HUMAN FACTORS

On the Moon's surface, the Apollo 14 astronauts Alan B. Shepard Jr. and Edgar D. Mitchell slowly worked their way toward the rim of a crater, dragging a small cart that bounced silently behind them. The view through the visors of their heavy space suits was stark and otherworldly, and the horizon seemed both close and far away. Science was their primary purpose, and they were committed to very ambitious rock gathering. Their flight followed close on the heels of Apollo 13, where an equipment failure had forced the crew to abort the Moon landing. Mitchell and Shepard were under intense pressure from NASA to make sure that their mission went flawlessly.[1]

A mile or so behind them, out of sight but not out of mind, was the tiny two-person lunar landing module. It looked unwieldy, flimsy, and not at all aerodynamic—whoever designed it must not have seen pictures of Buck Rogers's spaceship. Yet this combination sanctuary and rescue vehicle contained everything necessary to sustain life for a few days. Its rocket motors were powerful enough to brake the descent to the Moon and for the "elevator ride" back up, but it was strong enough only because of the Moon's low gravity. Outside it stood an American flag hanging limply in place.

Just a few miles above, the third member of the Apollo 14 team, Stuart Roosa, orbited the Moon in the command module, the bell-shaped vehicle with the elongated nose that took the astronauts to the Moon and back. Except for a few brief periods of blackout (such as when the command module was on the other side of the Moon and during a brief period of extremely high temperatures while reentering

Earth's atmosphere) the astronauts were linked to one another and with mission control by radio. Inside the command module was some of the most advanced equipment available in 1971: an array of dials and controls that might have bewildered the uninitiated eye; automated systems that sustained life and performed simple, routine tasks; and devices that stood guard to warn of unseen dangers. By today's standards, much of this technology would seem almost primitive.

At the Mission Control Center in Houston, Texas, scores of operators sat before banks of lights, gauges, and video screens, monitoring every phase of the mission. Linked to one another by a then state-of-the-art voice communication system, these experts relied on telemetered data, voice transmissions, and television images. At times, they accessed data banks or consulted with experts scattered across the globe. Sometimes they threw switches, adjusted knobs, or used computer keyboards to activate equipment. Not that far away, carefully trained NASA spokespeople briefed reporters, who in turn relayed word of the astronauts' progress. Although attention was focused on Mitchell, Shepard, and Roosa, they were but the tip of the iceberg that was Apollo 14.

SYSTEMS

Apollo and other space missions are large, complex, multidimensional entities that I will call systems. Like many other systems, Apollo 14 encompassed people (the ground-support crew, orbiting crew, and surface crew), tools (the command and landing modules, computers, radios), and activities. While we should never lose sight of the fact that systems are integrated wholes, they can still be "decomposed," or broken down, for purposes of analysis. The system's components—for example, the pilot orbiting in the command module, and the crew on the lunar surface—are themselves systems or "subsystems." It's all a matter of perspective.

Useful for us is Elwyn Edwards's SHEL model.[2] This straightforward formulation originated in aviation psychology and is intuitively compelling in our age of computers. SHEL is an acronym for four system components: software, hardware, environment, and liveware. Software is the procedures or practices that are in use; it is found in the policies, rules, regulations, standards, and customs that impart predictability and consistency to human behavior. Hardware, of course, is apparatus or equipment and in space might include propulsion, navigation, life support, and communications systems. Environment is the setting in which the activity occurs; for example, at mission control, onboard a spacecraft, or on the lunar surface itself. Finally, liveware is the number and types of crewmembers onboard, their abilities and limitations, their motives and preferences, and their attitudes toward their work and one another.

The SHEL formulation offers us four basic ways to improve systems. If we work with software, we can structure situations and tasks so that people can perform efficiently and effectively, with a minimum of mutual interference. If we work with hardware, we can develop better spacecraft, habitats, equipment, and tools, and we can improve the kinds of supplies stowed aboard. If we focus on the environment, we can create better working conditions by ensuring that the spacecraft or habitat is neither too hot nor too cold, nor too crowded, cramped, shaky, and noisy to work in efficiently. Finally, to improve liveware, we can find or select people who have the requisite physical, mental, and emotional talents and then increase their suitability through training.

Human Factors

The term *human factors* encompasses all of the conditions that affect human performance. Human factors experts draw on many fields to identify and then create conditions for optimum performance. Their primary goal is to devise better systems, ones that are relatively economical and efficient, do the job right, and do not break down. I will use the term *spaceflight human factors* to designate all of the conditions that affect human performance in space, as well as efforts to take these conditions into account while planning for and executing space missions.

In the narrow sense, human factors is rooted in biology and engineering. It focuses on the individual and his or her job. It accommodates human anatomy, physiology, biomechanics, and cognition (information processing) through the design of workplaces, the development of tools and equipment, and the organization, sequencing, and pacing of tasks. In the broader sense, notes Philip R. Harris, human factors also includes psychological, sociological, political, economic, and cultural variables, and encompasses people's emotions, attitudes, personalities, and interpersonal behaviors.[3] In this broader sense, human factors takes into account aesthetics, mental health, group and organizational dynamics, society, and culture. And, in this broader sense, human factors experts seek improvements in all three spheres of life: work, self-maintenance, and play. It is human factors in this broader sense that concerns us here.

There are strong theoretical and practical justifications for understanding spaceflight human factors, broadly defined. Spaceflight offers us a unique laboratory for expanding our knowledge of human adaptive processes. By conducting studies of human behavior in space, we increase the scientific dividends of our space efforts. More practically, a thorough understanding of human behavior in space has important implications for spaceflight operations. We want to protect spacefarers from an accumulation of stresses that could lead to psychiatric breakdown, high levels of interpersonal strife, significant performance lapses, or for that matter, even minor performance lapses at critical times.

Many of the central issues were set forth in an influential monograph prepared by the psychiatrist Nick Kanas and his associate W. E. Fedderson in 1971.[4] With some variations, these issues have come up again and again. The message is a simple one: the psychological, sociological, and cultural dimensions of spaceflight are very important and they have yet to be adequately addressed. Even the occasional astronaut such as Norman Thagard openly acknowledges this.[5] In 1987, a report of the Space Studies Board of the National Research Council concluded that "there is reason to suppose that psychological problems have already occurred in spaceflights and that these will increase in frequency" (given the demands of future missions).[6] Referring back to this report, a later report issued by the Space Studies Board in 1998 noted, "In contrast to the routine collection of data on cardiovascular, neurological and musculoskeletal changes in flight since the early days of the US space program, there has been relatively little effort to collect data on behavior and performance in a systematic fashion."[7] The issues identified in the earlier report were in fact extremely important, but over the intervening years NASA took few steps to address them.

In the mid-1990s we learned that NASA officials not only ignored such topics, they actively discouraged researchers from pursuing them. This is detailed graphically in a book by the former NASA psychiatrist Patricia A. Santy, *Choosing the Right Stuff: The Psychological Selection of Astronauts and Cosmonauts*.[8] She reports termination of promising behavioral research projects, destruction of valuable records, suppression of reports, and silencing of dissenting researchers.

Why might NASA be reluctant to address issues that could be critical for mission success?[9] Perhaps mission planners and managers simply are not aware of human factors or of their importance. Perhaps the "hard" scientists who control the program and who are used to accurate quantitative results find the "soft" sciences (such as anthropology and psychology) fuzzy, imprecise, and somewhat untrustworthy. No amount of argument can obscure the fact that the physical (and, to a lesser extent, biological) sciences seem able to provide "right answers" to questions while anthropologists, psychologists, and sociologists hedge their bets.

Human factors research is not necessarily in the immediate best short-term interests of those who plan, manage, or participate in space missions. Project managers who sponsor human factors studies may be put in uncomfortable positions if the results somehow reflect poorly on their management skills. They also run the risk of undercutting their own discretion if the human factors expert offers a recommendation that runs counter to the managers' wishes. Spacefarers who allow themselves to be scrutinized risk not only the imposition of new rules in formerly discretionary areas but also possible disqualification. This could come about if they reveal some sort of personal limitation or shortcoming, or if they say something that displeases their boss.

But perhaps the two greatest barriers to powerful and effective research into spaceflight human factors are the "untarnished astronauts" and the "too many cooks" problems. One of many strategies that NASA used in order to sell the public on the idea of space exploration was to present the astronauts as flawless individuals who represented the highest ideals of 1960s America.[10] This image served many purposes, including directing attention away from cost overruns and technical difficulties in the early space program. To preserve this image, NASA sought to protect the astronauts from close scrutiny. The faintest hints of incompetence, emotional instability, or flagging motivation could disqualify an astronaut, and, worse yet, decrease funding for the space program.

NASA, notes Howard McCurdy, sought to control the astronauts' images as tightly as a movie studio might control the image of a movie star. The press was very much a part of the "cover-up"—not of mental illness or criminal behavior, but of the minor flaws that make us human. Reporters from *Life* magazine, who had special access to the original astronauts, knew of shaky marriages, infidelity, and drinking but never reported these because NASA wouldn't tolerate it. Reporters who strayed too far from NASA's version of reality risked losing their accreditation.[11]

Given that NASA was trying to understate the risks of spaceflight and protect the astronauts' image, it is not surprising that they would be less than receptive to researchers who might ferret out astronauts' psychological problems or make dire predictions about how they would fare in space. At that time (and perhaps today) the risks of a tarnished image and reduced funding outweighed the risks of a human malfunction in space.

In order to keep programs alive, NASA has to satisfy numerous constituencies: the president, Congress, the Office of Management and Budget, major aerospace contractors, interest groups, and the military. A program that NASA wanted to be driven by its views of proper technology is compromised as the space agency evades budget cuts, finds cosponsors, and enlists international cooperation.[12] Each of these constituencies has its own requirements, such as keeping costs down or making sure that a spaceship's hold is large enough to carry military payloads. Eventually, "too many cooks spoil the broth" as the project is modified so that everyone finds it acceptable (even if no one finds it exciting). Thus, the International Space Station under construction bears but a slight resemblance to the huge, multipurpose affair that NASA had envisioned initially. By minimizing human factors inputs, NASA reduces the number of demands placed upon the project. Perceived costs and delays outweigh the desire to provide the strongest feasible on-site support for our spacefarers.

In stark contrast to NASA, Russia has had ongoing interest in spaceflight human factors. As one cosmonaut bragged, "We Russians, we know far more about psychological issues than the Americans; even the Americans admit this."[13] It is not coincidental that in the joint United States–Russian publication *Space Biology and Medicine,* the only two chapters relating to psychology are written by Vyascheslav I. Myasnikov, Iltuzar S. Zamaletdinov, Yuri A. Aleksandrovskiy, and Mikhail A. Novikov![14]

Given NASA's stiff resistance to spaceflight human factors, why is it that behavioral scientists have been integral to the Russian space program? Mary Connors observes that Russians have tended to have longer duration flights than Americans.[15] Longer flights, she notes, amplify the need for understanding human factors because some conditions that can be tolerated in the short term cannot be tolerated for extended periods of time. She adds that important cultural differences may be at play. Traditionally, the Russians have maintained a communal world view that emphasizes the needs of the group, while Americans have stressed the value of personal initiative and competition. Given their interest in social harmony, it is not surprising that the Russians took more steps to ensure good social relationships in outer space.

The Changing Conditions of Spaceflight

The 1970s marked an era of transition for NASA. The heady days of the 1960s with their almost unlimited funds, widespread public enthusiasm, and crash programs gave way to restricted budgets, tempered enthusiasm, and a rethinking of the next steps for establishing a continuing human presence in space. The key to maintaining a human presence in space, affordable only through cooperation with the U.S. Air Force, was the National Space Transportation System, or space shuttle. This reusable spaceplane was expected to ferry astronauts to Skylab, a U.S. space station that orbited earth during the early 1970s. The shuttle and space station missions were to differ in many ways from earlier flights—these craft were to carry relatively large crews consisting of men and women from many different fields. It was time to rethink all aspects of spaceflight, including the psychological and social dimensions.

How could this be done at no risk to the astronauts and NASA? The answer was simple—authorize a review of published literature but don't involve actual spacefarers in the project. Sponsored by R. Mark Patten and undertaken by Mary M. Connors of NASA Ames Research Center, Faren R. Akins, and myself, our work appeared in 1985 as *Living Aloft: Human Requirements for Extended Spaceflight.*[16] Our review centered on three changing conditions of spaceflight: crew size, crew composition, and mission duration. Later we added a fourth variable: although spaceflight has always been a high-tech operation, it is becoming increasingly so, making it important to ask new questions about the fit between people and technology.[17]

First there was a tendency toward increased crew size. Solitary spacefarers gave way to groups of two and three, and the present-day shuttles have carried crews as large as eight. On the one hand, increased size means more people to do the work. More talents and skills can be brought to bear on a problem, and there is more allowance for backup. Moreover, the presence of additional spacefarers brings welcome sources of social stimulation and diversion. Yet, increasing crew size raises new questions about motivation, coordination, and leadership. As groups get larger, each individual member may feel less responsible for the mission's success. The more people there are, the more difficult it is to organize and coordinate them, and there is an increased risk of problems due to factions and cliques.

Changes in crew composition or makeup were and are in the direction of increased diversity or heterogeneity. Initially, spaceflight was the province of white male military test pilots. Today, mixed-gender and international crews are common, and despite the *Challenger* explosion, which imposed a hiatus on efforts to introduce "civilians" into space, we can expect continuation of this trend toward greater diversity.

Like increased size, increased diversity promises both benefits and challenges. Not the least of the benefits is the fact that it opens opportunity to more people and reinforces democratic ideals. In addition to increasing the range of talents and skills and enhancing the richness of social interaction, broadening the base of participation also encourages greater "buy in." Indeed, widespread popular support for space activities may await programs that average people believe could include "someone like me."[18] The challenges of increased diversity flow from greater physical and cultural variability. Including people of different sizes and physiques, who have different languages and cultural preferences, raises many questions about interpersonal relationships.

The third change that prompted *Living Aloft* was increased mission duration. Rapidly, in the 1960s, space exploration moved from missions involving hours to flights involving days and then weeks. Although shuttles can accommodate large crews they are intended for brief flights, and only under emergency conditions can they stay in orbit over two weeks. However, in the early 1970s, the Skylab missions maintained crews for as long as eighty-four days. Since that time, cosmonauts aboard Russian space stations such as Mir set record after record, and at least one was aloft for over a year.

In theory it could take astronauts far less time to reach orbit (eight minutes) than to drive to work. Once they arrive at a space station, however, they may be expected to stay there for three or four months. It is very expensive to transport crews into orbit and back, so keeping crew rotation to a minimum cuts costs.

The conditions that were acceptable for brief forays into space are not, however, acceptable for living and working in space for months at a time. A simple analogy: harsh camping conditions, which are even fun on vacation, wear thin over time, and most people would not consider these conditions acceptable for their daily lives. The transition from visiting space to living in space occurs at about one month.

The fourth and final trend is burgeoning technology. There have been massive technological advances since the dawn of the space age, especially in such areas as computers, automation, and robotics. Many of the computers used in missions as late as the 1970s were of meager capacity by present-day standards. Some of these computers, impressive at the time, had on the order of 16 kilobytes of memory, only one-four-thousandth of the memory of many inexpensive home computers just twenty years hence (64 megabytes). Such technology—especially intelligent technology that senses, analyzes, and responds—must be understood fully as we plan new generations of space missions.

Lessons from Space, Lessons from Earth

To understand human adaptation to space, we look to the biological and social sciences. Unfortunately, all but the tiniest fraction of observations that make up these fields were made on Earth. Adaptation is a function of the person and the environment. How you do depends on both *who* you are (your abilities, personality traits, attitudes, and other predispositions) and *where* you are (the nature of the physical setting, the other people present, and the task at hand). The conditions that we are accustomed to on Earth are not the same ones we encounter in space. This means that simply observing people under normal, everyday conditions isn't good enough.

There are four conspicuous ways that spaceflight environments differ from everyday environments. First, spaceflight environments are dangerous. Spacefarers may be blown up along with their rockets, killed from the buildup of excessive force in the course of acceleration or deceleration, or perhaps even shaken to death. Spacefarers can be roasted or frozen, receive fatal doses of radiation, or run out of air. At any point in the voyage—from liftoff to touchdown—equipment failure or human error can kill.

Second, people who enter outer space become isolated. They have to make do with available supplies. They are cut off from their families and friends. The normal resources and activities that most of us take for granted are not accessible to them. Friendships are placed on hold. Relatively few other people are encountered. There may be a sense of being left out; for example, of being prevented from taking part in important family events or participating in major cultural celebrations such as May Day or Christmas.

Third, spaceflight environments are confined environments. Spacefarers are cooped up with one another in a habitat that isn't all that big. They encounter the same people, the same things, the same activities day after day, and unless an individual participates in a demanding and exhausting space walk or other extravehicular activity, there is no opportunity for escape.

Finally, at this point in world history, people who venture into space are deprived of most of the amenities of everyday life. Because of the expense of lifting materials from Earth, and because of small, cramped habitats with minimal storage space, supplies are necessarily limited. Certain items, such as fresh foods, may be impossible to obtain. To understand the human side of spaceflight, then, we need to understand people in dangerous, isolated, and confined settings that have limited supplies and amenities. Thus, we can learn from studying people in space, in spacecraft simulators or mock-ups, and in places like nuclear submarines or polar regions, which are characterized by isolation, confinement, and risk.

Spaceflight

Almost forty years have elapsed since the first astronauts and cosmonauts entered space. True, relatively few people have been in space at any one time, but over the past four decades around four hundred spacefarers have logged tens, perhaps hundreds of thousands of hours in space. "Long term flights of US and Soviet (Russian) space crews on the Skylab, Salyut and Mir stations," note Myasnikov and Zamaletdinov, "have provided an enormous amount of factual material relating to the behavior, mental health, performance, and interpersonal relationships that define the lives of crews in flight."[19]

We can and do draw on the observations of spacefarers, but in comparison to what one might expect given forty years of opportunities our knowledge is uneven and limited. We have learned a fair amount about biological adaptation, but beyond this we are forced to rely on astronauts' and cosmonauts' anecdotes and the results of the occasional survey. Anecdotes are informative but are of limited value; we need systematic studies so we can learn what is likely to happen in the typical case. It is interesting to learn about Shannon, Michael, Sergei, or Sasha, but for planning future missions it is even more helpful to know about spacefarers in general. This requires systematic quantitative studies of astronauts and cosmonauts, and it is this kind of studies that we sorely lack.

Simulated Spaceflight Environments

Another strategy for gaining needed information is to simulate the conditions of interest and then see how people react to them. This could involve constructing a mock-up of a spaceship or habitat and then having people do the kinds of things that we want them to do in space. As Harvey Wichman points out, a simulation is something like a play.[20] The mock-up comprises the stage setting and props. The crew is the cast, and their activities are prompted by some sort of script. This may result in a low-fidelity simulation, akin to when a cast first sits around in a circle and reads the script out loud. Or it may result in a high-fidelity simulation, akin to a full dress rehearsal.

During the 1950s and 1960s, space-cabin simulation studies mimicked, in certain ways, environments and situations expected during the early space program.[21] Astronauts have always trained in mock-ups, including shuttle simulators that react to the pilot's use of controls in the same way that airplane cockpit simulators respond realistically during make-believe airplane flights. There are several space station mock-ups, and we can expect the best of these to be among the training grounds for tomorrow's astronauts.

Simulations give us useful information, but as Jack Stuster points out, they also have severe limitations.[22] Unless we are willing to invest phenomenal amounts of money and energy, we can simulate only key parts of a mission (such as launches and landings). A realistic simulation of a multiperson mission lasting weeks or months and involving all operational phases (launch, transit, surface exploration, and so forth) would be enormously expensive. Another serious drawback is that in certain key areas simulators lack authenticity. We need to know how people behave in physically and emotionally challenging environments fraught with deprivation and danger, but ethics prevent us from putting people at high risk simply for the purpose of collecting behavioral data. Furthermore, Stuster points out that in some simulation studies people have simply quit and walked out before the study was completed, an option that would not be available during a real space-flight.

Another strategy is to study people in spaceflight-analogous environments. These are naturally occurring settings that involve danger, isolation, confinement, deprivation, and perhaps other elements that we associate with life in space. Whereas relatively few people have ventured into space, many people have lived in environments that are in some way reminiscent of space, and, over the years, we have made progress understanding how people adapt to them.[23]

Isolated groups of workers at weather stations, missile silos, and even remote national parks are worthy of study. We can look also at people who test their endurance by mountain climbing, living in caves, or taking intercontinental balloon rides. Prisoners in solitary confinement may seem promising, but early on NASA excluded this population because the people we hurl into jail are so different from the people we hurl into space. Despite many options, maritime environments and polar environments are the most frequently studied stand-ins for outer space.

Maritime Environments

For centuries, sailors have worked on and under the seas. Several periods of seafaring have informed discussions of life in space. The first of these was the age of exploration during the fifteenth and sixteenth centuries, when adventurers set forth from Europe to explore the rest of the world. These intrepid sailors weathered heavy seas in small, cramped ships and made do with poor rations. They were pioneers in every sense of the word and did not know what to expect. To some of them, sailing off the edge of the world was a serious prospect. Many of these sailors were cut off from their native lands for years at a time, and some of them never did return home.

Then, in the eighteenth and nineteenth centuries came the era of whaling ships. These magnificent ships left port on journeys that would end only when their barrels were full of oil. Voyages could take two years or more. Some sailors spent almost all of their adolescent and adult lives on these ships. In some cases, the charges they ran up in the ship's stores exceeded their share of the proceeds, hence they had to enlist for the next voyage in order to pay down their growing debt.

Conditions had improved by the time of the whalers. The ships themselves were bigger and better, the seas were charted, and captains knew where to go for resupply. Yet working the sails in a high wind and heavy seas remained formidable. There was great danger in killing a whale, stripping the blubber while standing on high, rickety platforms, and negotiating a pitching and gyrating, oily wood deck. Sometimes the fires that melted the blubber got out of control. Food had not improved all that much from the age of exploration—maggoty biscuits and rotten meat remained staple fare. The superb steaks that could have been cut from the whale's carcass were not considered suitable for human consumption.

The anthropologist Ben Finney describes the "second age of exploration," when sailing ships set forth from Europe to explore the South Sea islands and other exotic lands.[24] These ships carried crews of sailors who had little formal education but who did have a deep understanding of the ways of the seas, and scientists, who had formal education but were not familiar with seafaring. On occasion, conflicts erupted between these two groups. Right now, of course, spacefarers have strong science backgrounds and understand the ways of the scientist. In the future, when spacefaring becomes routine, there may be a large gulf between spacefaring scientists and spacefarers from other occupations, such as food service or construction work.

Early sailing ships are imperfect analogues to spacecraft. There are tremendous discontinuities between then and now. Unlike seafarers in the age of sailing, spacefarers will not have much flexibility in terms of their destinations, nor are they likely to stumble across unexpected islands that offer resupply. Unlike early seafarers, who traded hardship at home for hardship at sea, spacefarers must make major sacrifices in their accustomed quality of life. Unlike the early seafarers, who were forced or sold into service, spacefarers are highly educated and compete for the privilege of participating.

Today, we might study "superships," especially supertankers that for all intents and purposes never pull into port. (The liquid cargo is delivered to and from the supertankers by means of pipelines that extent far out into the harbor.) These extremely heavy vessels are unaffected by all but the largest waves. Yet they are very dangerous, since a spark in the wrong place can blow them apart, and when they are overcome by a storm they may break in two and sink. Sailors aboard such ships live in relative comfort in a highly automated environment under monotonous conditions. Superships may be useful analogues for interplanetary freighters.

In the early 1990s, Reid Stowe proposed taking his homemade sailing ship, *Anne,* on a special voyage to help us learn how people might get along with one another on a three-year trip to Mars.[25] Noting that his ship, a two-masted auxiliary powered, steel-and-fiberglass schooner, was a demanding vessel, he proposed tailoring the voyage to approximate Mars flight conditions as closely as possible. The *Anne,* twelve meters long and displacing sixty tons, would set forth with a crew of six to eight (the same size as an initial Mars crew under some scenarios) and three years' worth of provisions. For a thousand days they would sail outside of normal trade routes and without entering port. The crew would include scientists who would study weather, water and atmospheric pollution, and ozone depletion in remote and little-documented regions of the world. Stowe hoped to conduct field tests of communications satellites, water purification systems, and other equipment potentially useful for exploring Mars.

That voyage would have several advantages over post hoc analyses of whaling trips and historical voyages of discovery. Compared to the crews of the whalers and other large sailing ships, the relatively small crew of the *Anne* would be closer to the right size. Because the voyage of the *Anne* and the Mars trip would take place in the same historical period, multicentury cultural and technological disparities would be reduced. Finally, it would be possible to apply present-day research methodologies to study psychological and social adjustment onboard the *Anne.*

The cost of this simulation would have been negligible compared to the cost of a Mars mission, but when all was said and done, just as in the case of a real Mars mission, it proved difficult to fund. However, Stowe and a small crew had kept records during a six-month cruise to Antarctica. On this earlier voyage of the *Anne,* the crew fought the lack of sensory stimulation with plastic filters that allowed people to bathe in different colored lights, and a "bag of tricks" that included scented herbs and spices, stones, religious artifacts, pebbles, sand, and other items that stimulated the senses and kindled fond memories of home. Some of the crewmembers were artists, and rather than feeling intimidated by isolation they found that it enhanced creativity.

Even more ambitious is Marshall Savage's proposal for a huge floating sea colony, Aquarius, to serve two purposes.[26] The first is to conduct sea farming, aquaculture, electrical power production, and other commercial activities to raise the money to put space habitats in orbit. The second purpose is to serve as a test bed and training ground for the people who would later live on these satellites. In Savage's view, the sea settlement's architecture, the high degree of isolation, the need to live a relatively simple material life (at least at first), and the need to get along with fellow settlers will make Aquarius useful for developing procedures that will smooth our course into space. These procedures range from political and legal systems to specific technologies, such as the processes for manufacturing healthy and tasty synthetic food.

More analogues are found under the seas. Of particular interest are the underseas research vessels of the 1960s and 1970s.[27] In these undersea camps with names like *Sealab* and *Tektite,* small groups of aquanauts lived crowded together for extended periods of time. These were very dangerous environments in that equipment failure or human error could have had fatal consequences. Coming aboard required a saturation dive, which meant that the aquanauts could not return to land without spending many hours in a decompression chamber. Otherwise, as they ascended to the surface, the rapid release of nitrogen into their bloodstream could have resulted in the bends, so called because it causes victims to double up in pain.

Although conditions varied depending on the specific habitat and mission, it was difficult to achieve good temperature and humidity control in such environments, and in some cases, the sanitary, sleeping, and eating arrangements were not very good. Aquanauts took turns donning their underwater gear and going outside to conduct scientific research, salvage sunken equipment, or do other work. The cylindrical shapes and tubular configurations of these underwater stations, coupled with conditions where some people worked or relaxed in the habitat while others wore special life support equipment and worked "outside," are highly evocative of a space station.[28] These underwater stations are gone now, but fortunately for us, useful behavioral research was conducted during their time.

Navy submarines offer additional opportunities for research. These ships range from tiny three-person subs to fleet ballistic submarines. Each of the latter has two crews of 16 officers and 148 enlisted men that take turns on successive sixty- to ninety-day missions. Once at sea, there is almost complete isolation from home and family because security considerations prevent the use of radio.

Despite all that we have learned about seafaring and submarine construction, and despite a great safety record, life is always dangerous under the sea. Onboard nuclear reactors and missiles increase the danger, as does the submarine's attractiveness to enemy forces in the event of hostilities. The possibility of contamination of the air filtration system weighs heavily on the minds of those associated with these ships; in fact, to avoid noxious fumes such as those given off by chlorine-based cleansers, submariners at one time polished brass fittings with hot pepper sauce. Unlike astronauts in outer space, notes the former navy psychologist Benjamin B. Weybrew, submariners in "inner space" are concerned about pressurization (which can occur as the immense pressures of the sea squeeze the submarine's hull) rather than depressurization, which can occur when the atmosphere inside a space habitat or space suit escapes into the vacuum of space.[29]

Compared to their diesel predecessors, today's submarines offer good air-filtration systems, high quality food, and acceptable living quarters. However, even though today's submarines are about twice the size of their diesel predecessors, added equipment and a larger crew translate to less space per person, so conditions are still cramped. For decades, Dr. Weybrew led a vigorous program of psychological research.[30] However, because the results could reflect military preparedness and compromise national security, many years passed before they were declassified.

Polar Environments

The "high latitudes"—Earth's polar regions—give us wonderful opportunities for learning about isolated people working under harsh conditions. For many human factors researchers there is a compelling resemblance between early camps at the North and the South Poles and on the Moon or Mars; between early explorers racing their sleds across ice fields and Apollo astronauts driving their lunar rover on the Moon; and between geologists and biological scientists conducting research in Antarctica and, someday, on Mars. Not all Antarctic locations are equally suitable. Some are more isolated than others, and, as Peter Suedfeld points out, some contemporary polar settings, while still isolated, are relatively spacious, comfortable, and well provisioned.[31] Both Devon Island in the Arctic and the dry plains in the Antarctic are seen as promising testing and training grounds for Mars.

Dale Andersen and his associates point out that while there is real danger in Antarctica, fatalities are rare.[32] This means that research or training could be conducted there with a low level of risk. Because a logistical infrastructure is already in place, we can take advantage of preexisting habitats and supply lines to study people there. Many of the scientists who work in Antarctica engage in the kinds of scientific activities that closely resemble the kinds we expect to occur on Mars. Finally, the Antarctic Treaty of 1961 is a proven framework for international cooperation in science. Therefore, Antarctica is a congenial setting for international behavioral-research projects leading to international missions to Mars.

Both polar and outer space environments are fraught with danger. During the long Antarctic winter, the average temperature hovers around −60°C and, especially when the windchill factor is taken into account, unprotected flesh can freeze instantly. There is always the risk of an equipment malfunction—for example, a failure of the bulldozers that collect ice to be melted for water. The terrain itself is rugged and foreboding, and people who venture out risk falling into a crevasse. At certain times there is very poor visibility, and under these conditions a person may have difficulty finding camp, even though it is nearby. The smaller bases are the most dangerous because they have less safety and medical equipment and because they have greater problems in such areas as rescue and resupply.

Engineering and economic considerations limit the size, design, and provisioning of some polar bases and all space habitats. If a dwelling has to be dragged across miles of ice by tractors, airlifted into the interior of Antarctica, or blasted into orbit, its construction must be compact if not cramped. Additionally, requirements for high structural integrity and life support systems devour interior space otherwise devoted to work, living, and recreational activities. For the same reasons, life may proceed without abundant equipment, provisions, or supplies.

Antarctica enables us to study groups of different sizes, ranging from large crews at main stations during the summer months to small crews at remote stations during the winter. Some of these crews remain on station for brief periods of time, while others remain for months, perhaps wintering-over in a location where they cannot be evacuated until the austral spring. We can conduct longitudinal studies as well, observing what happens as a base grows in size.

As Chris McKay points out, Antarctica gives us an excellent model for "continuous presence."[33] Although any individual tour of duty is limited, successive tours constitute a series of interconnected, overlapping missions. Some stations in Antarctica have been continuously inhabited by a succession of crews for fifty years, a testimony to our ability to maintain human presence in hostile environments.

Polar human-factors research could be of particular benefit to the Mars mission. There is still time for a carefully planned, large-scale research program before the first humans leave for Mars. Arctic or Antarctic facilities allow us to test structures, supplies, and equipment, as well as people, a fact that has not gone unnoticed by U.S. aerospace and architectural firms. Of interest to both polar and space enthusiasts are small, compact, transportable research stations. Other areas of mutual interest are the equipment that will perform reliably under trying conditions, and training programs for those who will operate and repair the equipment. Indeed, robot explorers used in Antarctica could evolve into another generation of explorers suitable for Mars.

If we can't find a suitable polar training base, we can build one. This would allow us to duplicate as many Mars-mission conditions as possible. The base could be geared to the size and composition of a Mars crew, and if research or training missions could not last as long as the entire Mars mission will, they could last as long as the first crew's stay on Mars. And, unlike in the case of space cabin simulators, we would not have to worry about trainees simply quitting and walking out the door. The Mars Society established a simulated training base on Devon Island in 1999.

CONCLUSION

It is people who design, construct, and operate systems for the exploration of space, and it is people who live and work in space. Even as we must understand the technical side of a mission, we must seek to understand the human dimension. To these ends, we draw on the field of human factors, broadly defined. We draw on a full range of biological, behavioral, and social sciences and seek the implications for easing our progress into space.

Both the U.S. and the Russian space programs have attended to human factors in the narrow sense, but the U.S. program has been slow to delve into such topics as personality, interpersonal relationships, and group dynamics. Psychological and social factors have always been important, and, as we look to the future, they will become increasingly so. Changes in crew size, crew composition, mission duration, and technology prompt us to rethink all aspects of spaceflight, including the psychological and social aspects.

The most obvious way to learn about psychological and social adaptation to space is to study spacefarers themselves. This is indeed useful, but so far relatively few people have flown. Another way to expand our knowledge is by simulating spaceflight conditions, but simulations can be very expensive, and, under most rules, participants can simply walk out. Still another possibility is to study people in environments that somehow resemble the environments we anticipate finding in space. Many environments capture some of the elements of life in space, but the two that have generated the most attention are maritime environments and polar environments.

We need to keep in mind, however, that spaceflight-analogous environments are but rough approximations of spaceflight environments. There may be points of correspondence, but there are dissimilarities as well. For example, the microgravity associated with the space station and the low gravity of Mars are not duplicated in Antarctica. Whereas at times only minimal protective clothing is required in Antarctica, this is not the case in space, where the unmodified atmosphere does not sustain human life. Due to the expense of lifting mass into orbit, even small polar bases seem spacious relative to initial off-world outposts. At smaller polar bases it is possible to have emergency camps or "safe havens" and to store five years' worth of supplies in case war or other events were to make Antarctica inaccessible. Spaceflight-analogous environments may be closer to spaceflight environments than the everyday environments familiar to you and me, but they are not identical. At this point in history, understanding the human side of spaceflight requires some guesswork, as well as painstaking assembly of information from many different sources.

CHAPTER 3

HAZARDS AND COUNTERMEASURES

To some observers, the flame billowing forth from the mighty Proton rocket had an orange tinge, suggesting that something was awry. But apprehensions were laid to rest that day in November 1998, when the Proton placed Zarya, the first module of the ISS, into orbit. Zarya serves as a fuel depot and control module. In early December, the space shuttle *Endeavour* carried the second module, Unity, into orbit and successfully connected it to Zarya. This required incredible skill and determination on the part of the astronauts. Over the next few years, completion of the 520-ton space station would require 12 Russian and 33 U.S. launches. Assembly would require 160 space walks and over 1,000 hours of extravehicular activity.

After noting that routine shuttle launches are about as exciting as watching "an 18-wheeler cruise Interstate 80," *Newsweek* reporters extolled the dangers of space station construction.[1] "It's not going to be pretty," said the NASA administrator Dan Goldin. "We're going to see some unbelievably tough problems." A NASA study estimated a 74 percent chance that one of the 45 launches could be lost. Fluctuations in temperature could cause Zarya or Unity to contract, expand, and burst apart. A nickel-size micrometeorite could rip a hole in a space suit, causing decompression so violent that the wearer's eyeballs would pop out. But the prospects of death seemed to breathe new life into America's space program.

Although a spacecraft may offer a shirtsleeve environment, outer space itself is harsh and unforgiving. Spaceships and habitats must protect occupants from wild temperature extremes, near-vacuums, and in some cases, poisonous atmospheres or corrosive dusts. Acceleration to orbital speeds, radiation, and floating under conditions of micro-gravity, or weightlessness, could have severe biomedical consequences. Thousands of pieces of space junk threaten life. These include spent boosters, discarded covers and mantles, dead satellites, and broken-up equipment ranging from large chunks to tiny pieces.[2] Landing back on Earth is dangerous, particularly if the spacefarers must be plucked from the ocean or rescued from the Siberian hinterlands. Brian Harvey observes that during the Cold War there was some risk of cosmonauts re-entering by parachute being mistaken for American spies. Fortunately, no cosmonauts were harmed by their compatriots.[3]

Space, for our purposes, refers to a region beyond Earth; when we enter space we move beyond the immediate influence of Earth and its atmosphere.[4] The first requirement for survival in space is withstanding the process of getting there. This requires riding atop a highly complicated and potentially explosive rocket at very high speeds. For the foreseeable future, space missions will be of two types: orbiting satellites and lunar or planetary surface exploration. In either case, people must live and work under conditions very different from Earth's.

We have evolved in such a way that we depend on Earth's atmosphere. Held in place like a shell around our planet, this atmosphere protects us and constitutes a nurturing medium that helps keep us alive. This shell is fairly tight fitting; as we proceed upward from sea level it becomes difficult to breathe at about 4,300 meters. As we proceed to 21,000 meters there is not enough atmosphere to provide lift for an airplane's wings or permit a jet engine to burn its fuel.

When we leave this protective shell we lose our best protection from meteors and comets, which usually break up at seventy to ninety kilometers above Earth and then incinerate as they progress downward into our atmosphere. We also lose protection from solar and other forms of radiation. We leave behind a cover that, accompanied by the warmth of Earth itself, moderates environmental temperature, despite our complaints about heat waves and cold spells.

Temperatures on moons and other planets depend on their distance from the Sun, the planet's composition and internal temperature, the extent to which the planetary surface absorbs or reflects heat, and, if the planet holds an atmosphere, its composition and density. In the absence of moderating conditions such as clouds, the transition from day to night may involve contrasts of hundreds of degrees.

On Earth, we evolved under gravitational forces that keep our feet down, and we developed a straight, upright posture. Our growth and development, our physiology, and to some extent, our neurological processes are fine-tuned for Earth's 1-G gravity. Gravity keeps us anchored, both physically and psychologically. Of course, we run, jump, play hopscotch, do somersaults, and in other ways defy gravity. We can withstand the effects of microgravity for a split second at the apex of a jump from a diving board, for twenty seconds or so during a parabolic (arcing) flight in an airplane, or for many months during an orbital mission. Nonetheless, we are best suited to a 1-G environment, and living in places where we seem much lighter or heavier has important implications.

Space is vast—the edge of the universe is perhaps 15 billion light-years away. Space is characterized by sparseness, not in the sense that there are not many stars and planets, but in the sense that they are separated by vast distances. Despite such distances, we have learned a lot about our neighbors in the solar system.[5]

Our closest natural neighbor, the Moon, is about 380,000 kilometers away. To put this in perspective, you would have to travel around the equator approximately nine and a half times in order to cover the distance from Earth to the Moon. With a diameter of 3,476 kilometers, it has about a quarter of Earth's diameter. It has a gravitational force that is 16 percent of Earth's, and there is no atmosphere on it. Each lunar day, equal to 27.32 Earth days, is almost as long as a terrestrial month. Temperatures range from 107°C during the lunar day to −153°C at night.

The lunar surface is quite variable and includes the smooth lava plains that typify the maria, or "oceans," and the rougher highlands. Because there is no atmosphere to burn up meteorites and no weather to lessen or erode the impacts, the Moon is pockmarked with craters. Edgar Mitchell mentions that one of the distractions that made it difficult for him to sleep during his stay on the Moon was a plinking sound suggestive of meteors hitting the lunar lander.[6] The Moon's surface is covered with a thick layer of coarse, highly porous rubble called regolith. Mountains are very tall but not very steep, so visitors returning from the Moon are likely to report rolling scenery rather than jagged peaks. If the Moon has metals they will be rare and concentrated far from the surface, near the center of the Moon's core. Researchers recently discovered frozen water in the shadows at the Moon's polar regions. Because of its proximity, and because astronauts have already visited it six times, the Moon will be one of the first sites for an extraterrestrial human community.

The first planet from the Sun is Mercury, with a diameter of 4,880 kilometers. Mercury has .37-G. Each of its days is the equivalent of 59 of our days. Like the Moon, Mercury has no atmosphere; average temperatures range from 350°C during the day (only 22°C above lead's melting point) to a cool −170°C at night. Because parts of Mercury are forever in the shade, it has temperature contrasts greater than any other known planet. Like the Moon's surface, Mercury's surface is highly cratered and covered with cracked lava. Mercury has a heavy core, something like Earth's, so at some point it might become attractive for mining.

Out beyond Mercury but still on Earth's sunny side is Venus. With .88-G, Venus is very similar to Earth, but each Venusian day, equal to 243 Earth days, is very long. Although the green planet was named after the goddess of beauty, during the twentieth century we discovered that it is one of the least hospitable known places for human life. Venus has a very thick atmosphere, which, measuring 90,000 millibars at sea level, is 90 times as dense as Earth's. (One thousand millibars, or 1 bar, is the pressure at sea level on Earth.) Venus is covered completely by clouds of carbon dioxide laced with sulfuric, hydrochloric, and hydrofluoric acids. The clear, green light that we see from Earth is sunlight reflected off of Venus's impenetrable cloud cover. The site of a runaway greenhouse effect, Venus has average surface temperatures reaching 480°C, and, because of the thick atmosphere, it never really cools. Although impenetrable to the eye, radar and other observations taught us that the surface is covered with plateaus and circular features, perhaps the result of major meteorite impacts. Much of what we know about Venus we learned from a series of Russian probes.[7] Venera spacecraft, landed on Venus by the Soviets in 1975, photographed rocks similar to those found on Earth.

Proceeding outward from the Sun past Earth we find Mars. With a diameter of 6,787 kilometers it is smaller than Earth, and at .38-G it exerts less than half of Earth's gravity. The length of the Martian day, which measures 25 Earth hours, is almost the same as the length of the terrestrial day.

Martian atmosphere consists of carbon dioxide, some nitrogen, and a small amount of other gasses, but the atmosphere is very thin—only

6 millibars at the surface, not even 1 percent of Earth's. Unlike on many other possible destinations, there are no triple-digit temperature swings—average temperature on Mars is $-23°C$ as compared to $22°C$ on Earth. Seasonal changes in the polar regions hint that water is available. Compared to the hellhole Venus and the barren Moon, Mars is a beckoning oasis, and it will be the first planet other than Earth to host human life.

As we proceed outward from Mars we could visit an asteroid belt and then Jupiter, Saturn, Uranus, Neptune, Pluto, and perhaps a remote tenth planet. However, the distance from the Sun to Mars (227×10^6 km) is only about a *third* of the distance to the next planet on, Jupiter (778.3×10^6 kilometers); and beyond that, as we move successively to Saturn (1427×10^6 km), Uranus (2870×10^6 km), and then Neptune (4497×10^6 km), the average distance from the Sun almost doubles.[8] Except for Pluto, the outer planets are giants, and whereas some of them may have promising moons their extremely cold, dense atmospheres and other features make them rather unappealing. For now, the Moon and Mars are daunting enough.

What known extraterrestrial destinations share is improvidence and lethality. Improvidence refers to their inability to offer us the essentials that we need in order to live. It is not possible to survive in these environments without bringing along not only clothing, food, and other supplies but also air to breathe and water to drink. Later on, it may be possible to develop local or indigenous resources, but now and in the near future these will have to be imported from Earth, at tremendous expense. Lethality refers to the fact that anyone who is foolish enough to enter these environments without sufficient protection will die from asphyxiation, freezing, roasting, exposure to radiation, or in any of countless other ways.

ENVIRONMENTAL RISKS

Some of the hazards that beset spacefarers are the same as those that beset people everywhere else, but they are magnified in space. These include accidents and injuries that are all the more dangerous because treatment facilities are inadequate and evacuation is impossible. Other hazards are more specific to outer space. These include the biomedical risks associated with acceleration, weightlessness, or microgravity, and radiation.

Acceleration

In the 1830s, people feared that the human body could not withstand the 40-kilometer-per-hour speeds of early railroad travel. About a hundred years later, scientists voiced similar concerns about pilots who sought to break the sound barrier. Today, notes G. Harry Stine, we know that humans are capable of traveling at almost any speed, providing they are protected from the wind and similar dangers.[9]

The problem is acceleration; that is, building up the speed necessary to enter orbit or cruise from one part of the solar system to another in a reasonable period of time. The faster the acceleration, the more profound the effect on the human body. For example, you wouldn't notice much if you gently pressed an automobile's gas pedal to reach 100 kilometers per hour over 45 seconds, but you would be pushed back in your seat if you "floored it" and attained the same speed in 4.5 seconds. Similarly, a slow, gradual stop would be barely noticeable, but the rapid deceleration of a panic stop would jerk you forward against your seat belt.

People can't tell a difference between the forces acting upon them due to gravity and those due to acceleration, so acceleration and deceleration can be expressed in terms of "G." Thus, a 2-G acceleration produces a force twice that of gravity, 3-G acceleration three times that of gravity, and so forth. Within limits, G effects are useful because they help us judge speed and direction, and accomplished pilots can react to even minor shifts in gravity before their flight instruments can respond.[10] Nonetheless, as acceleration and G-forces increase, not only do we feel progressively heavier, but we find it more difficult to move our arms and legs, and then our wrists and fingers. Circulation is impaired, consciousness dims, and there is the possibility of blackout. Very high G-levels have bad consequences: truly compacted bladders, lungs that can't breathe, hearts that can no longer pump blood.

Acceleration and deceleration were of great concern during the first years of spaceflight, but are of less concern now. During these flights, which involved military rockets and parachute landings, spacefarers were exposed momentarily to as much as 11-G. Shuttle riders are not subjected to more than 3.0-G. However, unlike the early riders of military rockets, today's spacefarers are exposed to these forces for up to seventeen minutes, so there are concerns about G effects on their weakened cardiovascular systems during landing.[11]

A person's ability to tolerate acceleration depends on posture and position relative to the direction of the acceleration. The countermeasure is to have the person's entire body face the direction of acceleration (or face away from the direction of deceleration) so that the forces are spread across the body rather than concentrated on specific areas, such as the top of the skull and the spine. Contour seating and composite materials also reduce stress and ease breathing and blood circulation. Seating provides extra protection for specific areas of the body (such as the lower back) and includes straps to keep the occupant from sliding around. Acceleration may reemerge as a major problem if we develop advanced propulsion systems capable of accelerating rapidly to interstellar speeds.

Microgravity

The unique biomedical risks of spaceflight are those associated with living under conditions of microgravity. Spacefarers experience a sense of weightlessness in orbit and during the cruise phases of lunar and interplanetary missions. These effects are not due to the actual absence of gravity, but to its cancellation during flight. Occupants of orbiting vehicles, for example, are in a continuous state of falling—falling around Earth.

Prior to the first space missions researchers had many worries about the effects of microgravity, most of which proved needless. Early flights allayed fears that the heart might not be able to pump blood effectively, that spacefarers couldn't swallow or digest food, or that their eyeballs might change shape, making it impossible to see clearly. Nonetheless, exposure to microgravity is accompanied by a collection of adverse effects that Leonard David has termed "space scurvy."[12] These include disorientation, difficulty maintaining balance, cardiovascular deconditioning, loss of muscle strength, demineralization of the bones, sleep disturbances, and other consequences. Some of these resemble the effects of aging, and, as a extra benefit, if we can understand the mechanisms that cause these problems in healthy young astronauts, we might be able to apply them to reverse the aging process and help people on Earth.

Shortly after entering microgravity, about half of all spacefarers experience symptoms reminiscent of car sickness or seasickness.[13] Space adaptation syndrome (SAS) symptoms include disorientation, pallor, malaise, loss of motivation, irritability, drowsiness, stomach awareness, and infrequent but sudden vomiting.

SAS is unpredictable since it is hard to guess in advance who will experience it, and a person who experiences it on one mission may not experience it during another. Incidence varies depending on mission profile—in general, the tinier the spaceship the less likely that space adaptation syndrome will occur. While none of the solitary astronauts who were crammed into the earliest U.S. capsules reported this malady, it affects somewhere between 60 and 70 percent of astronauts on their first shuttle flights.[14] Symptoms tend to peak during the first two days, and recovery is complete by the third or fourth day, although symptoms have persisted longer than this on some Russian missions. In one study of eighty-five shuttle astronauts, fifty-seven reported SAS. Of these fifty-seven cases, 46 percent were considered mild, 35 percent moderate, and 11 percent severe.[15] Compared to men, women were somewhat less affected, and seasoned spacefarers are less likely to fall victim than beginners.

Although the symptoms are not entirely incapacitating and tend to disappear in the first few days, they are dangerous if they peak when spacefarers have to undertake difficult maneuvers. For this reason, shuttle flights last a minimum of three days (to ensure that everyone has recovered prior to undertaking delicate landing maneuvers), and other demanding activities such as docking and space walks do not take place until after the third day of the mission, when symptoms have disappeared.

One way or another, theories of SAS involve the physiological effects of weightlessness. Specialized nerve endings throughout our bodies called receptors bring us the information that we need to orient ourselves and adapt to our environments. Some of these receptors—for example, rods and cones in the eye—detect light and import information from the external world. Other receptors located deep within our skins bring us information about our physiological states, posture, positioning, and movement relative to gravity. From early childhood we effortlessly transform these different sources of information into coherent wholes, and in the process become oriented toward our surroundings.

Neurological patterns ingrained in us on Earth are interrupted in microgravity. No longer are we anchored by our sense of up and down; we may float in almost any position relative to the interior of the habitat, and, if we look out the window, we might be surprised by our position relative to Earth's horizon. Fluids shift within the head and the otoliths (tiny mechanisms within the inner ear that provide us with a sense of orientation and balance) no longer send us a familiar pattern of signals. The information coming from our eyes and from our balance mechanisms no longer match.[16]

Attempts to weed out people who are susceptible to SAS have not progressed very far because the condition is not all that predictable. Other solutions involve training people to avoid or withstand the symptoms. Some researchers try to toughen people by exposing them to disorienting situations on Earth, hoping that they will develop a resistance that will serve them well in space. Other investigators use a technique called biofeedback, which involves training people to control their bodily responses to disorienting conditions. Some of the medicines used to combat air sickness or seasickness help, and these are the most frequently used medicines in space, but we don't want spacefarers to have their senses dulled as they undertake maneuvers that require focused attention, clear thought, and a steady hand.[17]

Spaceflight leads to predictable changes in our muscles. These adaptations to weightlessness contribute to performance decrements in space and may have adverse consequences when the spacefarer returns to the normal gravity of Earth. John B. Charles, Michael W. Bungo, and G. William Fortner trace changes in cardiopulmonary function under microgravity to fluid shifts within the body.[18] On Earth, gravitational forces cause blood to move downward and pool in the lower part of the body. Within a day or so after a person enters space, this pooling no longer continues, and fluids shift upward toward the head. Visible symptoms include a certain puffiness of the eyelids and face. The body interprets this pooling as excess fluid. In the course of eliminating this fluid, blood plasma decreases by about 12 percent and total body water by about 2–3 percent. This fluid shift, which begins almost at once, stabilizes in about five days. Although this is a proper response to microgravity, the body must readapt when it is again subjected to 1-G. The fluid immediately shifts downward, in essence draining the blood from the heart and the brain. This leads to symptoms of low blood pressure, including weak pulse, rapid heartbeat, dizziness, and the possibility of fainting.

Countermeasures to cardiac deconditioning include vigorous exercise and use of the lower body negative-pressure device, a piece of equipment that is reminiscent of a large, sacklike blood pressure cuff that can be pumped up to redistribute fluids in the way that they would be distributed in normal gravity. Another technique is to increase body fluid shortly before deorbiting. Shuttle astronauts consume generous quantities of fluids accompanied by salt tablets to aid retention, thereby helping to restore fluids to prespaceflight levels.

On Earth, our muscles are gravity fighters. They allow us to maintain our distinctive erect, bipedal posture. Furthermore, when we walk, jog, or run we must push ourselves upward as well as forward, and whenever we lift something with our arms we must counteract the gravitational forces. Under microgravity, we do not have to counteract gravity. The problem with this life of ease is that we get out of shape. Disused muscles become weak muscles and don't work as well as when they are fit. This becomes evident when the spacefarer leaves the habitat to help construct a space station or partake in some other vigorous activity, and upon return to Earth. Following a 211-day flight on the Russian space station Salyut 7 in 1983, cosmonauts Anatoli Berezovoy and Valentin Lebedev returned so debilitated that they were barely able to walk for a week, and required intensive rehabilitation.[19]

Although there are wide individual differences, skeletal muscular deconditioning is evident in as little as five days.[20] On brief missions, the deconditioning probably reflects the reduction in load due to microgravity. On longer flights there are morphological changes to the muscles and muscle fibers. Muscles diminish in size, lose maximum strength, and become less resistant to fatigue. There may be muscular twitching and loss of fine motor control. These effects seem to be progressive, and we have no firm basis for forecasting their effects beyond a few months. Thus, the extended periods of microgravity that will accompany the Mars mission are a source of concern.[21]

Measurements taken during bed-rest studies as well as in space itself reveal that a reduced load on the skeleton leads to changes in bone metabolism and the elimination of calcium through urination and defecation.[22] A study of Skylab IV astronauts revealed that the amount of calcium passed with urine increased until thirty days into the mission and then reached a plateau, while the amount eliminated with feces continued to rise over the course of the eighty-four-day mission. Although initially slow, calcium loss increased to three hundred milligrams per day by the end of the flight.[23] Other studies confirm a progressive loss of calcium and bone mass over time, a change associated with a weakening of the bones and resulting in increased risk of fractures and other skeletal problems.[24] As calcium leaves the bloodstream it is processed in the kidneys, where it can contribute to the formation of kidney stones, which are no fun under any circumstances, never mind when one is isolated from medical treatment on Earth.

Finally, microgravity, perhaps in association with the other stresses of spaceflight, may affect the body's ability to defend itself against injury and illness.[25] After entering microgravity there is a fairly rapid loss of plasma volume and red blood mass. Less reliably, there is a reduction in red blood cell count, and other changes in blood chemistry that have possible health implications. Analyses of white blood cells and serum proteins show that space induces changes in the immune system. This means that in space, wounds may take longer to heal and illnesses may be more difficult to shake because of immunosuppression. Although some of these changes reverse shortly after return to Earth, spacefarers may be particularly susceptible to illness during an initial postflight period.

Exercise helps counter many of the adverse effects of microgravity, including cardiovascular and muscular deconditioning and bone demineralization. From Skylab on, spacefarers who followed special exercise regimens have done better than control subjects who did not. Today, there is a vast array of exercise equipment that is designed for space. All of these devices are compact and lightweight, and they work under conditions of microgravity. Thus, chest expanders, spring-loaded rowing machines, treadmills, and stationary bicycles that use friction to create load are useful, but a set of barbells would never do. A recent review by Charles F. Sawin suggests that although spacefarers on extended-duration missions exercise as much as two to four hours per day, even lesser amounts help. Twenty minutes of moderate to intense exercise (60 percent to 80 percent of preflight maximum workload) three times a week helps spacefarers preserve their strength, and treadmill exercise helps maintain neuromuscular patterns required for walking and running.[26]

Ted Wade and his associates propose an automated fitness management system, or FMS.[27] This would rely on a data bank that included each spacefarer's medical record and activity preferences, along with information about their work assignments. The FMS would monitor fitness changes automatically and plot fitness trajectories. It would then draw upon a broad array of options such as diet, exercise, anabolic drugs, and electrical stimulation of the muscles to develop a tailor-made plan for each spacefarer. As the spacefarer followed the customized plan, performance would be carefully monitored, and the spacefarer would receive detailed feedback. This holistic approach is intended to reduce the time that crewmembers would spend on exercise machines.

Radiation

In space, people are bombarded by potentially significant doses of naturally occurring radiation.[28] Some of this comes from the Sun, leading some nuclear power advocates to quip "a day without radiation is like a day without sunshine." This radiation streaks toward Earth, too, but relatively little actually reaches us.

We are to some extent protected by the magnetosphere that extends some 65,000 kilometers and traps radiation. The Van Allen belts, located in the magnetosphere where protons and electrons are trapped and accelerated to high energy, turn relatively benign particles into dangerous radiation, meaning that this could be a very poor place to linger. The magnetosphere provides in effect an umbrella of protection for spacefarers in low Earth orbit, as well as for people on Earth. Earth's atmosphere offers additional safety for those of us on the ground. The thicker the atmosphere, the greater the protection, so people who live on mountaintops receive slightly more radiation than people who live on the seashore.

Spacefarers are exposed to radiation from three different sources. First are cosmic rays, most of which originate outside of our galaxy and form an ever-present background of highly energetic charged particles. Cosmic rays are simply atomic nuclei accelerated to very high speeds, perhaps attaining 80 percent of the speed of light. Their strength varies over time and is more concentrated in some parts of the universe than in others.

Second, when our Sun flares, there are solar particle events, another source of radiation. The Sun flares continuously, but only the largest outbursts produce dangerous levels of radiation at Earth and beyond. Although quite variable in intensity, there are cyclical variations, with peaks every eleven years. We have some ability to predict these flares, but only on a short-term basis—a few hours, not days or weeks, in advance. (We can predict these storms because, while we can see the flares at the speed of light, the protons travel a little slower and take longer to reach us.) There were solar storms of lethal intensity during February 1956, November 1960, and August 1972, but fortunately nobody was on their way to the Moon at those times.[29]

Finally, there is radiation from the nuclear fuel and other substances that we bring with us into space. Despite deep-seated fears of radiation and environmentalists' protests, the fact remains that properly handled nuclear power may have great value for future operations. Nonetheless, as in all other applications, radioactive sources can be improperly stored or mishandled, and pose a hazard to human health.

Cosmic rays, solar particle events, and artificially created radiation are not identical, nor do they have identical effects on the human body. Suffice it to say that we are justifiably concerned about their combined effects on spacefarers, but that it is very difficult to make precise predictions.[30] We have to rely on mathematical models, simulations, animal studies, and clinical observations of people who were victims of rare industrial accidents or affected by atomic blasts.

Within wide latitudes, the effects of radiation, like those of other pollutants and poisons, are probabilistic and depend on the intensity and duration of the exposure, the initial physical condition of the victims, and the degree of protection available to them. Like small glasses of red wine, very low levels of radiation may have positive effects. Slightly larger doses result in no apparent damage to a person. Either the person isn't affected, or the body quickly repairs itself.

Under most conditions, the immediate effects of radiation poisoning are less worrisome than the results of long-term cumulative exposure. Radiation causes bone, blood, and other cells to malfunction and die. Radiation can cause cataracts, initiate benign and malignant tumors, and alter the genetic code, leading to infertility or stillbirth. Lymph tissue, bone marrow, and gastrointestinal linings are particularly sensitive to radiation, while the central nervous system, muscles, bones, and connective tissues are resistant.[31]

We can protect some spacefarers from radiation by bringing them back from Earth orbit if we believe that the cumulative dose is getting too high, or we can limit exposure by not letting anyone stay there for more than a set period of time; for example, ninety days in the case of the International Space Station. Because of concerns over the cumulative effects of radiation, we may not want to train spacefarers on the ISS before they set off for an extended stay on the Moon or Mars. We can develop safe havens; that is, heavily shielded areas or "storm cellars" where people can retreat in the case of major solar flares or when passing through radiation trapped and accelerated within the Van Allen belts. Given that we can detect solar storms only a few hours in advance, we could not retrieve people from space before the storms hit, but we can give them a relatively safe place to hide. The problem is that the heavier the shielding the greater the protection, but the higher the cost of getting it into orbit.

Spacefarers who reach planetary surfaces will find additional protection.[32] Keep in mind that in outer space, radiation comes from all directions. Any barrier between oneself and the heavens reduces exposure, so if you were to lie on your back on the Moon, the radiation that would otherwise enter through your back would be shielded by the Moon's mass. A habitat that is situated next to a cliff would be protected on the cliff side as well as underneath, and a habitat that is nestled under a ledge would have some overhead protection as well.

The Moon has no atmosphere, but Mars's thin atmosphere offers at least some protection from radiation. In both places, it will be possible to cover habitats with powdery, rocky soil (regolith) and reduce risk from radiation in much the way that we might reduce risks in underground nuclear fallout shelters.

In addition to "hardening" habitats we can "harden" the spacefarers by using techniques designed to keep soldiers as fit as possible in the course of a nuclear engagement.[33] Preventative measures include a healthy diet with plenty of green vegetables, coupled with massive supplementary doses of vitamins A and E. At the time of exposure, risk can be reduced further by atropine injections. Together, these two measures can reduce the effects of radiation by as much as 40 percent. Finally, there are various types of postexposure therapies that purge toxins from the body and help repair the immune system.

Maintaining Health in Space

Missions are jeopardized when crewmembers become incapacitated and cannot perform their jobs properly, or when their inability to deal with illness forces mission control to abort the mission or to attempt an expensive and risky medical evacuation. Consequently, we seek to minimize the likelihood of injury or accident in space and find prompt, effective treatments that allow spacefarers to stabilize and then resume their duties as soon as possible.

In addition to equipping ourselves to deal with the medical hazards of space itself—space adaptation syndrome, cardiopulmonary deconditioning, mineral loss, radiation poisoning—we must deal with everyday medical problems that just happened to occur in space. Even highly trained people working in supremely engineered environments can have accidents—such as cuts, burns, sprains, and broken bones. Furthermore, spacefaring does not grant immunity from normal illnesses—heart attacks, appendicitis, gallstones, toothaches, constipation, itchy scalp, common colds, bacterial infections, and viruses.

Two broad types of measures—preventative and remedial—promote wellness in space. Preventative measures take place on the ground prior to departure and in space. These are intended to minimize the number of health-related problems that arise. Remedial measures take place in space and are intended to stabilize, ameliorate, or cure health problems whose first symptoms appear in space.

Preventative Measures

Astronaut selection procedures constitute an early and very strong line of defense against health problems in space.[34] Although the medical standards have relaxed somewhat since the earliest years of spaceflight, medical examinations still eliminate people who are likely to have conditions that will impair their performance or that are inconsequential on Earth but could become troublesome in orbit. After selection, spacefarers undergo a series of physical examinations and run the risk of medical disqualification almost to liftoff.

Good health is important, but the superbly fit, highly tuned athlete may not be the best choice for microgravity. People who place a premium on fitness may undergo greater deconditioning and be more bothered by it than people who are simply in good health. This is suggested by studies showing that there is little advantage to rigorous preflight exercise: that under many conditions athletes' work capacity decreases more rapidly than that of nonathletes; and that older, less physically active individuals adapt better to some of the stresses of microgravity than do their younger, more active counterparts.[35]

A second line of defense is quarantine or restricting astronauts' contact with other people so that they do not come down with a communicable disease that will manifest itself after takeoff. In the earliest days of spaceflight, colds and flu were not considered much of a problem; the missions were so short that if an illness hadn't broken out by the time of departure it was unlikely to become full blown until after the astronauts had returned to Earth just a few hours later. For longer missions, various quarantine procedures were tried, with varying degrees of success. Although today's astronauts do not undergo a strict quarantine, the number of people they encounter is restricted.

A third line of defense is keeping the spacecraft as immaculate as possible. This means attending to personal hygiene, proper handling of food, and correct maintenance and frequent cleaning of sanitary facilities. Unfortunately, as Rosalind Grymes and her associates point out, despite heroic efforts to sanitize spacecraft a variety of microorganisms are likely to lurk onboard. There, in the closed-loop life support of the space vehicle, these microorganisms will be passed from one person to another and perhaps themselves recycled.[36]

In-flight prevention includes monitoring environmental systems, maintaining good nutrition, and using exercise facilities to counteract the deconditioning associated with microgravity. In later discussions of life support systems, habitability, and off-duty activities I will take a closer look at some of these measures.

In-Flight Medical Treatment

In-flight treatment requires trained medical personnel, medical supplies and facilities, and support from physicians on Earth. An obvious alternative is to include a physician in the crew. Besides giving medical care, a physician could conduct biomedical research and perform other scientific duties. Russians do include physicians on their missions but only occasionally are they assigned to shuttle flights. Since there are many jobs to be done in space and not that many people to do them, it is efficient to find ways to give the crew the services of a medical practitioner without adding another person to the crew.

NASA trains astronauts in the effects of spaceflight conditions on human physiology and psychology and in handling illness and injuries in space.[37] All shuttle astronauts are taught cardiopulmonary resuscitation and first aid. Two members of each crew are designated crew medical officers, or CMOs. These astronauts have additional medical training, maintain medical supplies and equipment, and monitor environmental hazards, as well as take the lead role in medical care. In flight, CMOs use the radio to work closely with medical experts on the ground.

During the astronauts' waking hours, a physician is on duty at mission control. While the astronauts sleep, doctors are about ten minutes away. Each astronaut is given a private medical consultation each day. The flight surgeons carefully monitor each astronaut's appearance and behavior. They listen carefully to the tone of the astronaut's voice. During particularly stressful periods, such as space walks, flight surgeons monitor telemetered data, such as from an ongoing electrocardiogram that would reveal heart arrhythmia or other problems.

Ground-based physicians can use television to diagnose specific illnesses. A small, portable unit known as the telemedicine instrument pack (TIP) includes a television camera with various lenses used for ear, nose, and throat; ophthalmological; and dermatological exams.[38] The TIP was field-tested in a rural clinic in Texas. In one set of tests, the patient and the diagnostician were across the room from one another, and in another set of tests the patient and the diagnostician were twenty miles apart. As the medical technician worked the compact camera, the diagnostician scrutinized images on a small, flat TV screen. The TIP was a little less efficient than an on-the-spot examination, and imperfections in the picture's color interfered with certain diagnoses. Nonetheless, TIP was useful for diagnosing many kinds of illness, including basal cell cancer, impetigo, nasal polyps, sinusitis, and diabetic retinal changes.

As spacefarers travel farther and farther from Earth, telemedicine will become less and less feasible, at least for emergency conditions. At great distances patients might not be able to survive long enough for a medical complaint to reach authorities on Earth and for the expert's advice to return to the party in space. Remote outposts will have to include one or more well-trained physicians among the staff, along with a database that can be used in lieu of hospital libraries.[39]

Medical supplies have been carried on all U.S. flights, although in the early days these included little more than a handful of pills. The choice of supplies is a difficult one: on the one hand, it is tempting to include medicine to cover as many contingencies as possible, but on the other hand weight and storage considerations limit the amount that can be carried. Shuttle flights include medical packs containing basic first aid equipment, pharmaceuticals, immobilization devices, and other means to deal with likely medical problems. A recent study of in-flight use of medicine based on 219 records (each one representing one person on one shuttle flight) showed that 94 percent used some medication. Of the astronauts who took medicine, 47 percent took remedies for SAS, 45 percent took medicine to alleviate sleep disturbances, and small numbers took remedies for headache, backache, and sinus congestion.[40]

The ISS will carry medical packs for everyday use and for special emergencies.[41] As new modules are added to the space station, and new crewmembers come onboard, additional medical capabilities will be added, perhaps culminating in a dedicated health maintenance facility. Future flights will require extensive provisions, ultimately including x-ray and surgical equipment and perhaps a convalescent area.

Some medical treatments are simplified under conditions of weightlessness, but others are made more complicated. In space, drugs may not be absorbed into the bloodstream, or take action, as occurs under normal gravity. This results from decreased bodily fluids, changed digestive processes, and altered metabolism. Drugs that can be administered orally on Earth may have to be administered intravenously or by means of a transdermal patch. Microgravity forces us to rethink both the type of prescription and the dosage.

Surgical procedures are complicated by weightlessness. Because of the closed environment, anesthetics that have to be inhaled are ruled out, and we cannot count on gravity to hold patients' organs and tissues in place. Moreover, like all other liquids, blood tends either to adhere to surfaces in very fine sheets or break up into increasingly small globules that could fill the cabin with a red mist. It will be difficult to keep surgical areas free of contamination and to keep surgery from contaminating the habitat.

Fortunately, in a young and healthy population surgery will be rare: among Polaris submariners there were 269 surgical cases in 7,650,000 man-days. In a continuously occupied space station, this would equate to a major surgical case about once every nine years.[42] Surgeons are working on devices (such as surgery containment chambers) that will make it possible to conduct surgery in space.[43] Perhaps the biggest problem right now is that we cannot expect CMOs to become trained surgeons. Until the distant future, surgical procedures are likely to include emergency appendectomies, wound suturing, and other measures intended to preserve life. Space stations are not sites for elective procedures such as plastic surgery or correcting hammertoes.

In a high-tech environment we expect high-tech medicine. Unfortunately, telerobotic surgery conducted by physicians on Earth doesn't seem promising, at least for long-distance missions accompanied by real-time communication delays. However, looking to the distant future, the NASA administrator Dan Goldin foresees chemical procedures that can heal people without scalpels or incisions, implanted monitors along with micromachines that can heal on command, and devices that deliver antibodies to designated problem areas.[44] By the time we go to Mars, spacefarers may be implanted with micro monitors and robots that will handle their medical needs.

As more and more people travel farther and farther from Earth we must be prepared to deal with both accidental and naturally occurring deaths in space. Robert M. Beattie recommends onboard facilities where autopsies can be performed to identify the causes of death, such as viral or bacterial infections that could threaten the rest of the crew.[45]

Such problems could come about if bacteria growing within the space-craft led to a variant of Legionnaires' disease or if explorers encountered a hostile extraterrestrial "bug."

Beattie adds that we must be prepared for storing or disposing of human remains and for assuaging the grief of remaining crewmembers. Space station plans call for stowing corpses onboard for return to Earth via the shuttle—there will be no burial in space reminiscent of burials at sea, because we do not want frozen corpses added to the mix of orbiting space junk. No one is pleased by the prospect of death in space, but as in the case of addressing other threats we are better off planning now than improvising later.

CONCLUSION

Around 1960, before anyone had actually flown in space, we had many fears but little information concerning how human bodies would react. Although studies of military pilots had given us considerable information on managing acceleration, we were apprehensive about the effects of microgravity. Would it be possible to breathe properly, swallow food, and eliminate bodily waste ? As a matter of fact, it is quite possible to do all of these things and a lot more.

Space adaptation syndrome is uncomfortable and annoying, but usually transient. The greatest risk is that the spacefarer who suffers from it will not be able to perform at an acceptable level, perhaps during some intricate maneuver early in the flight. SAS remains hard to predict, and whereas certain treatments seem to benefit certain spacefarers on certain missions, there is no sure way to keep the malady at bay. Yet it has not kept us chained to planet Earth.

The redistribution of fluids under microgravity, the resulting deconditioning of the cardiopulmonary system, the loss of skeletal muscle mass and strength, and a weakening of bones as a result of mineral depletion are all legitimate concerns. We do not want people to become dizzy, faint, or lose strength under any conditions, never mind when lives depend upon their performance. Here again there are no "magic bullets," but a combination of measures, including exercising, consuming dietary supplements, donning special pressure devices, and loading up on fluids and salt prior to return to Earth, have managed to pull spacefarers through. Such regimens have led to a decoupling of physical deconditioning and length of time in space.

Except for large, rare, and at least somewhat predictable solar flares and passage through the Van Allen belts, radiation levels within our solar system are not likely to produce acute effects. However, we do worry about chronic effects, including cancer, cataracts, infertility, and stillbirth. We can reduce the risk by retrieving people from orbit, helping them predict and hide from solar flares, and having them use dietary supplements and medicines to build resistance to radiation.

Despite early worries, through a combination of protective and corrective measures we have been able to manage human health in space. It is true that such problems as deconditioning and radiation will intensify as missions become longer and longer. Yet we are moving gradually into space, feeling our way, and discovering new countermeasures as we go.

CHAPTER 4

LIFE SUPPORT

Some of the diamond mines near Johannesburg, South Africa, reach three thousand meters underground, and the deeper they are the higher their interior temperatures. Most are very humid, and conditions in these artificial tropics border on the intolerable. Heat stroke is a risk. As the miner's temperature rises, symptoms include headache, irritability, delirium, and finally irreversible brain damage. As A. W. Sloan points out, we have two ways of coping with such extreme conditions.[1] Physiological adaptations are the bodily adjustments we make as we strive to maintain equilibrium. Upon entering the mine, body temperature, cardiac output, and pulmonary ventilation rise. Sweating is profuse and becomes dilute, lowering the rate of loss of vital body salts.

Behavioral adaptations are the choices we make to prevent or cope with the harsh conditions or their consequences. Miners should drink plenty of water and take extra vitamin C. Their diet should be low calorie, relatively high in carbohydrates, and relatively low in protein and fats. Clothing should be loose, light, and porous. In the mines, air conditioning is impractical, but huge fans force air through them. Each hour, the workers' temperatures are taken, so that those who become too hot can be evacuated at once. People on the verge of heatstroke can be saved by packing them in ice or, if ice is not available, by covering them with light cloths that are sprayed with water and fanned. In the worst parts of the mines, workers wear frozen vests covered with insulated jackets.

Our uniquely powerful and effective brains, our ability to communicate with one another, and our technology have given us means to spread into almost all ecological niches on Earth. These same characteristics make it possible for us to occupy niches off our planet. We bring with us, into the near vacuum of space, artificial atmospheres that allow us to breathe, heating and cooling systems that moderate vast temperature swings, adequate if not copious quantities of water and food, and everything else that we need to survive.

SPACECRAFT AND HABITATS

Space habitats are the places in which people live in space. There are two types: nonplanetary (spacecraft, satellites) and planetary. They must be built right and built to last despite incredibly harsh environmental conditions. We cannot risk failures or set the stage for an endless parade of makeshift repairs.

Engineering and economic considerations limit the size, design, and provisioning of space habitats, and there are necessary trade-offs between cost and comfort. Mass is of overwhelming importance. In the early days, engineers started with the maximum weight that their rockets could put into orbit, and then worked backward to determine the habitat's size.[2] Dwellings that have to be blasted into orbit must be compact, although through modular construction large space stations are possible. Requirements for high structural integrity, propulsion systems, and life support systems add mass and diminish interior space that could be devoted to work, living, and recreation. Until we can tap better the immense amounts of power available in space we must steer away from design features that require a lot of electricity.

Designers have a limited choice of materials. Flight-qualified materials must not outgas, that is, vent noxious fumes. Many materials vent chemicals that, in high concentration, are harmful. Literally hundreds of such chemicals were found within the confines of early U.S. spacecraft. That great "new car" smell that so many of us love emanates from such chemicals as formaldehyde, but these chemicals do not make us sick because we do not stay inside our cars day after day. Materials must be fire resistant and easy to clean. If the temperature-humidity balance isn't quite right, habitat surfaces, like shower curtains, become congenial hosts for unpleasant forms of microbial life.

Although different generations of spacecraft may bear a resemblance to one another, spacecraft have continually evolved over time. They have grown in size and complexity, and the life support systems have become better at making occupants comfortable. Let us briefly review the evolution of American and Russian craft here, as a prelude to tracing the development of life support systems for use in space.[3]

Visiting Space and the Race to the Moon

During the Cold War of the 1950s, the United States had encircled the Soviet Union with military bases and intermediate-range ballistics missiles (IRBMs). Because the Soviets did not have comparable bases in North America, their defense depended on intercontinental ballistic missiles (ICBMs) that could be launched from Soviet territory to targets like Boston and New York. By 1957 the Russians had developed a reliable, powerful rocket that outstripped anything in the U.S. inventory. This put them in an excellent position to start the space race, and it began with a flourish when the Soviets launched the first Sputnik in October 1957.[4]

A young Soviet Air Force officer, Yuri Gagarin, was the first human to orbit Earth; he lifted off at 9:07 A.M., Moscow time, on April 12, 1961. His Vostok (meaning East) capsule consisted of two sections, which included an instrument module and a 2.5-meter-diameter re-entry vehicle. There were three portholes, and Gagarin used a small, free-floating doll as a gravity indicator. The Vostok orbited approximately 150–350 kilometers above Earth.

Because animal flights had suggested that humans might have difficulties adjusting to space, Gagarin was limited to one circle around the world. There was some doubt about the Vostok's retrorockets' ability to kick the spacecraft out of orbit for its return to Earth. Just in case, his spacecraft was provisioned for ten days in space, enough time for natural orbital decay to bring him back if the rockets failed. The retrorockets worked, and the final part of Gagarin's descent was achieved by using an ejection seat that shot him out of the capsule to make a soft parachute landing, while the capsule made a harsh landing on its own. His flight lasted 118 minutes.

There were five more Vostok missions before the program ended in 1963. On one of these, the cosmonaut Valery Bykovsky remained in space for 5 days and 81 orbits, a record for solo spaceflight that remains today. Before he landed, another Vostok mission was under way, carrying Valentina Tereshkova, the first woman to fly in space. At one point, Bykovsky and Tereshkova passed within four and a half kilometers of one another.

Voskhod (Sunrise), a modified version of the Vostok, flew in 1964–65. It carried the first 2- and 3-person crews into space. Like all Russian spacecraft that would follow, the flights were equipped with both a parachute and rockets that slowed final descent, allowing cosmonauts to remain onboard until touchdown. There were only two crewed flights, and the 3-person flight was so cramped that not only did the engineers have to remove safety and life support equipment, the cosmonauts had to diet to reduce overall launch weight. An interim spacecraft, the maneuverable Voskhod enabled the first extravehicular activity, or "space walk."

Russia's Soyuz (Union), an all-new design, carried its first cosmonaut in 1967. It consisted of 3 modules—a service module, an orbital module, and a descent module. Measuring 9.14 meters long by 10.05 meters wide (including solar panels) Soyuz could carry as many as 3 passengers. Soyuz has undergone substantial evolution: between 1967 and 1970, Soyuz saw regular service as an orbiting spacecraft, and in 1970 it was modified to serve as a transport vehicle to the first Russian space stations. The most recent generation, Soyuz-TM, first flew in 1987. Today, the Soyuz TM model is still in use, and ferries spacefarers and supplies to Russia's Mir Space Station. This highly reliable vehicle and its successor will make regular runs to the ISS.

Sputnik was a tremendous propaganda as well as scientific coup for the Soviet Union, and the United States was caught unawares. Americans had not been under the same pressure to develop an ICBM as had the Soviets, and the responsibility for U.S. rocket development was spread across three military services. The U.S. space program was initiated and turned over to civilian authorities by President Dwight D. Eisenhower in 1958, and three years later President John F. Kennedy proposed that Americans land on the Moon within the decade. Three series of flights—Mercury (1961–63), Gemini (1965–66), and Apollo (1967–75) got us there. Mercury tested America's ability to put people in space, keep them alive, and retrieve them. Gemini established that 2-person crews could work together, undertake space walks, and connect with other orbiting vehicles. Apollo missions included final tests of the people, hardware, and procedures; moved on to swings around the Moon; and then completed the actual landings.

These three sets of expendable spacecraft bore a superficial resemblance to one another (they all looked something like gumdrops with cylindrical noses) but differed vastly in size, sophistication, and accommodation. Flights ended when the spacecraft left orbit and parachuted back to Earth. The occupants were cushioned by a relatively soft splashdown in the ocean, where they remained until rescued by sailors.

Mercury vehicles were tiny capsules, little more than enclosures for the individual astronauts who flew in them. About 3 meters long and under 1,400 kilograms, they orbited 160–288 kilometers above Earth. There were 23 Mercury launches: 16 were successful, and 7 failed; fortunately, none of the failed flights carried passengers. The first two piloted flights, carrying Alan B. Shepard Jr. on May 5, 1961, and Virgil I. "Gus" Grissom on July 21 of the same year were suborbital, conducted in part to test systems and procedures and in part to join the Russians in space as soon as possible. There were 4 astronaut-bearing orbital missions, beginning with John H. Glenn Jr. on February 20, 1962, and concluding with Gordon Cooper Jr.'s 22-orbit, 34-hour mission.

Gemini is Latin for *twins,* and Gemini craft, almost 6 meters long and 3 meters in diameter and weighing in at just over 3,800 kilograms, carried a crew of 2. Some Gemini astronauts took part in extravehicular activities, or space walks, to hook up with docking targets launched on Agena rockets. There were 10 piloted Gemini flights in 1965 and 1966, including Frank Borman and James A. Lovell Jr.'s 14-day mission that removed all doubts about human capabilities to live in space long enough to visit the Moon.

Much larger than their predecessors, the Apollo mooncraft consisted of 3 modules. The command module carried 3 astronauts and contained the flight controls. The service module housed the life support and propulsion systems for use after the huge Saturn V booster had completed its work. The command and service modules, which formed the moonship, were a little over 10 meters long and had a maximum diameter of just under 3 meters. The third module was the lunar module, or LM. A flimsy contraption that carried its own rockets, the LM ferried 2 passengers between the lunar orbit and the lunar surface. It could be flimsy because it did not have to withstand Earth's atmosphere during flight and because the Moon's low gravity made for easy landing and takeoff.

Three astronauts rode from Earth to the Moon in the command module. There, one astronaut remained onboard, circling the Moon, while the other two used the LM to descend to the Moon's surface. After they had completed their work the pair would use the LM for the "elevator ride" to rejoin their orbiting companion for the trip back to Earth.

Just before Christmas 1968, astronauts Frank Borman, James A. Lovell, and William A. Anders left Earth, swung out around the Moon, circled it ten times, and then returned. On Sunday, July 20, 1969, the Apollo astronauts, Neil Armstrong and Buzz Aldrin, set foot on the Moon while pilot Michael Collins remained in lunar orbit. Among other things, Armstrong and Aldrin ate hot dogs, drank coffee, and collected 21 kilograms of Moon rocks. They left behind an American flag, scientific equipment to conduct automated research, and assorted equipment that was jettisoned to reduce weight for the elevator ride back up. All told, there would be 6 successful landings and 12 astronauts would roam the lunar surface. The Apollo program was a success, but the last people to visit the Moon returned home in late 1972.

Shuttles and Space Stations

America's space shuttle, officially called the Space Transportation System, or STS, is a reusable vehicle that can land on a runway like an airplane and consists of 3 parts.[5] These are the giant external tank that carries the immense amounts of liquid fuel needed to help lift the shuttle to orbit, the solid rocket boosters that provide additional lift, and the orbiter proper. It is the orbiter, which has an airplanelike appearance, that actually circles Earth. After liftoff the spent boosters are dumped into the sea and then salvaged.

The orbiter is reminiscent of a commercial airliner, the DC-9, but its habitable volume is only 71 cubic meters. The front of the orbiter includes the flight deck on the top level, living quarters on the middle level, and life support control equipment on the lower level. The midsection has a generous payload bay 18 meters in length. In its basic form the shuttle can carry as many as 8 people for 16 days, but in an emergency one or two more can be packed aboard. The latest version, *Endeavour,* can remain aloft for up to a month. A shuttle's payload bay sometimes carries special-purpose modules, such as the European Space Agency's Spacelab, which until 1998 supported research in the physical and life sciences, or the U.S. Space Hab, which extends the spacecraft's habitable volume.[6]

Designed in the 1970s, *Columbia,* the first shuttle to fly in orbit, did so with a crew of 2 on April 12, 1981, coincidentally the twentieth anniversary of Yuri Gagarin's debut in space. During its early years of operation, the shuttle would orbit Earth while the occupants conducted scientific research, tested commercial processes thought to benefit from microgravity, and undertook odd jobs such as placing or repairing communications satellites. In the mid-1990s the shuttle began work as a ferry, first to Russia's Mir Space Station and now to the International Space Station. The Russian *Buran* is a reusable shuttle highly reminiscent of the U.S. shuttle but was flown only once and then without a crew. It was put in storage and then scrapped after the dissolution of the Soviet Union.[7]

Space stations are facilities launched without a crew, but they are occupied by a series of crews who arrive on successive transit vehicles. They are intended to remain in operation for extended periods of time. Skylab was the first—and thus far only—exclusively U.S. space station. It was built from unused components of a Saturn rocket of the type used to launch Apollo missions. Compared to earlier and contemporary spacecraft, Skylab was enormous, and it set new standards for accommodations. Its life support systems were the basis for today's shuttle and ISS systems. At 38 meters long and with a primary living space 14 meters long and almost 7 meters in diameter, it was aptly dubbed "a house in space" by science writer H. S. F. Cooper.[8] It hosted 3 crews in 1973–74. These three missions lasted 28, 59, and 84 days and heralded the transition from visiting space to living in space.

In the early 1970s, the Russians began launching a series of Salyut (Salute) space stations.[9] These evolved into successively more sophisticated and habitable dwellings, culminating in Salyut 7. The first crew that tried to reach Salyut 1 in 1971 failed to dock properly, and the three disappointed cosmonauts had to return to Earth. The second crew successfully docked and orbited for 24 days, but as they returned to Earth, a faulty valve leaked air and they were asphyxiated. Salyut 2 and Salyut 3 experienced technical difficulties; only Salyut 3 was occupied, and it did not remain functional for very long.

In January 1975, everything came together with Salyut 4. This station afforded, among other luxuries, Velcro carpeting, an exercise machine, and a shower. The first two crews orbited for 28 and 60 days, respectively. Success continued with Salyut 5, launched in 1976, and Salyut 6, launched in 1977. Unlike its predecessors, Salyut 6 had 2 docking ports, making it possible to add Soyuz transit vehicles at each end of the station, enabling departing crews to stay onboard until the incoming crew had arrived, and allowing robot supply ships (called Progress) to support long-duration missions. Salyut 6 and its successors initiated a string of international missions that at first involved guest cosmonauts drawn from Soviet bloc countries and later from western Europe, Japan, and the United States.

Mir (Peace) replaced the last Salyut in 1986. An evolutionary rather than radical step forward, Mir is distinguished by 6 docking ports that allow the station to expand its capabilities with docked modules. Transporting scientific equipment and supplies on separate modules means that the modules do not have to be stored on Mir at liftoff. Thus, less internal space has to be used for storage, and more can be devoted to living. Modules provide additional storage and living space and carry specialized research equipment such as spectroscopes and smelting furnaces to permit experiments with metals. Mir is about the size of five or six school busses arranged in Tinkertoy formation, and with its huge solar arrays it is reminiscent of a dragonfly.[10]

From the early 1970s until the late 1990s Russia's space stations built the foundation for future space missions. During these years the Russians perfected docking techniques and became proficient at resupply missions. They took photographs of Earth; conducted astronomical, biological, and other types of research; explored the effects of prolonged weightlessness; and continued to develop exercises and other countermeasures to combat deconditioning.[11]

As early as the late 1960s NASA had plans for a 100-person, permanent space station. Skylab, a part of the Apollo Applications Program, was conceived in 1967–68 as a logical follow-on to Moon flights using surplus Apollo and Saturn hardware. NASA continued to plan for a successor to Skylab, but the project was plagued by tremendous cost overruns, and, in the early 1990s, President Bill Clinton's administration canceled the project and urged that it be redeveloped as an international venture. The Russians, in no position to launch a successor to Mir, agreed, hence the International Space Station.

The leading partners are the United States and Russia, with strong contributions from Canada, the European Space Agency nations, Japan, and several other countries. The project is overseen by the U.S. aerospace giant Boeing. Participating nations help share the cost, and each has its assignment for producing specific, identifiable components, so that each has a visible stake in its success.[12]

The ISS will be the largest space structure by far, measuring 88 meters long, 110 meters wide, and weighing almost 420,000 kilograms. It will orbit at an average altitude of 407 kilometers. The first module was the Zarya (Sunrise) block developed by the Russians and put into orbit in late 1998. Construction will take five or more years to complete.

We are not sure what types of spacecraft will take us back to the Moon and carry us to Mars. However, they are likely to bear enough of a resemblance to past spacecraft (such as Apollo) that we can make intelligent guesses about the human factors issues that will be involved.

LIFE SUPPORT SYSTEMS

Penelope Boston recommends four basic requirements for life support systems.[13] First, they must be appropriate for the mission and the objectives. On early missions it made sense to simply bring drinking water along, but for the Mars mission it will be necessary to recycle water in transit and find or produce water at the planetary base. Second, life support systems must be highly reliable. A broken air conditioner on Earth is easily fixed, but a broken cooling system in space could force evacuation of a crew. Third, life support systems must not require a lot of a maintenance or even attention from the crew. Spacefarers should be spared from routine housekeeping activities so that they can devote their time to scientific and commercial work. Finally, while we can discuss life support systems individually, they should work together as integrated, coordinated wholes. For instance, the amount of water that can be carried or produced onboard has implications for waste management, and attempts to grow crops could change the gaseous composition of the habitat's atmosphere.

Artificial Atmosphere

On Earth, uncontaminated air consists of approximately 20 percent oxygen, 78 percent nitrogen, and small amounts of other gasses and substances (even tiny amounts of airborne substances can have powerful effects). Atmospheric pressure at sea level is 1 bar (1,000 millibars) which equates to 14.7 pounds per square inch. On Apollo-era flights, atmospheric pressure was kept low, in the vicinity of 350 millibars or 5 pounds per square inch, but in contemporary shuttle flights it duplicates the atmosphere at sea level.

Some relatively brief space missions have used almost pure oxygen, while others have used a mixture of oxygen and nitrogen. There are important trade-offs here. Unlike a mixture of oxygen and nitrogen, pure oxygen has the advantage that a sudden decrease in cabin or space suit pressure will not automatically trigger decompression sickness. If nitrogen is present in the bloodstream, decompression forms bubbles that cause incapacitating and perhaps fatal cramps. Breathing pure oxygen purges nitrogen from the system, and decompression sickness cannot occur. A pure oxygen environment, however, is itself poisonous if people live within it for extended periods of time. In pure oxygen, materials combust rapidly if not explosively, so there is a greater risk of fire.

Early U.S. missions involved pure oxygen. On the shuttle, astronauts breathe a normal oxygen-nitrogen mix but switch to oxygen for a three-hour period preceding extravehicular activity. This purges the body of nitrogen prior to opening the air hatch and reduces the risk of the bends if there were to be a failure of the air lock or the space suit somehow were to decompress. On contemporary missions, filters and purification systems cleanse the air. Air quality monitors alert the crew to oxygen depletion and the buildup of toxic substances due to out-gassing or microbial infestations.

We will not be able to bring along enough oxygen and other gasses to satisfy the requirements of settlers on the Moon or Mars. The Moon is completely devoid of an atmosphere and thus is particularly challenging. Under one scenario, hydrogen will be transported to the Moon and then used partly to produce additional water and partly to help extract oxygen from the powdery, topmost layer of lunar regolith.[14] After the regolith is scraped from the surface, oxygen-bearing ilmenite will be separated from the less useful material (which constitutes about 90 percent of the mass) by means of established electrostatic techniques. Next, the separated ilmenite will be fed into a hydrogen reduction unit, where it will be heated under pressure and about 10 percent of the mass will be extracted as oxygen, which will then be combined with the hydrogen as water vapor. The process won't be cheap, because production requires heavy equipment and plenty of electricity. Eventually it might be possible to get the necessary electric power from the Sun, and lunar oxygen or "lunox" could be a major source of oxygen used by people throughout our solar system.

Although thin and not already suitable for human life, Mars's atmosphere contains oxygen, buffer gasses including argon and nitrogen, and water. Thomas R. Meyer and Christopher P. McKay describe autonomous processes that will extract and stockpile oxygen from Mars's carbon dioxide atmosphere for use by the first settlers.[15] Over the long haul, other processes could increase the level of breathable oxygen in the Martian atmosphere itself.

Temperature

We have many techniques to protect ourselves from temperature extremes in space. In transit, spacefarers use the "barbecue roll," which involves a slow rotation of the craft so that portions of the spaceship take turns baking in the Sun and keeping cool in the shade. This keeps any one part from becoming superheated. Skylab used a giant "awning" to shield the habitat from the direct rays of the Sun. We can also use thermal insulation, which on the Moon or Mars may consist of rocky soil stacked on top of the habitat. Spacefarers there may protect themselves from cold climates (and radiation) by living in habitats reminiscent of Midwestern underground houses with sod roofs. Heating and air conditioning will be available but more difficult to engineer than on Earth, in part because down here we capitalize on airflow partially controlled by gravity. The shuttle has radiators in its bay; after the bay doors are opened, interior heat dissipates into space.

Spacewalkers are sometimes exposed to direct sunlight for extended periods of time. The work itself is strenuous, so heat generated by metabolic processes is added to the heat of the Sun. On the Moon's surface, temperature was further increased when the astronauts' space suits were covered with a dark, heat-absorbing dust. Space suits are ingeniously designed to keep the heat out and have powerful air conditioning, but the prolonged periods of heavy labor required to construct the ISS could overload space suit air-conditioning units and increase risk of heatstroke.

Water

People in arid regions have learned to get by with relatively small amounts of water, as have residents of polar and outer space environments, where freshwater is in short supply. In many locations in Antarctica, despite the fact that snow is abundant it is not all that easy to harvest and melt, so rationing is necessary. Thus there are no extravagant "Hollywood showers." Instead of having a leisurely soak, people wet themselves down, lather up, turn off the taps while they scrub themselves, and finish with a brief rinse.

Despite many economies, spacefarers will still require a lot of water to get by. Penelope Boston reviewed estimates made between 1968 and 1987 suggesting that drinking-water requirements will run between 3.6 kilograms and 4.6 kilograms and wash water requirements will be between 1.5 kilograms and 18 kilograms, for a total consumption of as much as 22.6 kilograms per person per day.[16] Another estimate suggests that supporting a crew of 6 on a 90-day mission would require 31,400 kilograms, or 31.4 cubic meters, of water.[17] It is extremely expensive to launch such massive stores.

On Mercury, Gemini, and Skylab, spacefarers simply brought water along. On Apollo and on the shuttle, water has been a useful by-product of the chemical reactions used to produce electricity by fuel cells. This water is pure and palatable and, on the shuttle, is used for drinking, food reconstitution, and personal hygiene. So much water was generated by the shuttle's batteries that it was possible for it to leave 635 kg of it behind after depositing an astronaut on Mir.[18]

Water reclamation and recycling are in use on Mir (about 60 percent is recycled) and will be essential on the ISS and on Moon and Mars missions. Recycling involves wringing humidity from the air and processing used wash water and even urine.[19] Interestingly, through breathing, sweating, and eliminating bodily waste people actually excrete *more* water than they take in—this is a quirk of human metabolism.[20] Recycled water isn't always potable, but not all water will be drunk; lower standards are acceptable for wash water.

Earth is a watery planet; the Moon and Mars are not watery. As late as 1996, the Moon was described as so dry that "if concrete were there we would mine it for water."[21] Recently researchers discovered ice at the Moon's poles, and there is some speculation that lots of water lies trapped inside minerals in the Moon's interior.[22] Mars is a dry planet, but evidence including ancient riverbeds suggests that water existed there at one time. Even though this may have been millions of years ago, the water could not have evaporated entirely. Thomas R. Meyer and Christopher P. McKay estimate that approximately 1.3 cubic kilometers is available in the Martian atmosphere.[23] This amount, they point out, would be enough to supply the needs of a 1,000-person base for 33,000 years. A combination of natural processes would replenish the water in the atmosphere over time.

Although there is no firm proof, significant amounts of water may be available from Mars's polar ice caps. The larger of these is 1,000 kilometers in diameter, and could be as much as 4–6 kilometers thick at certain spots. The other is only 750 kilometers in diameter and may reach a maximum thickness of 1–2 kilometers. Other possibilities include a permafrost rich in water. Perhaps this could be liberated by melting the permafrost and extracting water from the mud. However, we are not yet in a position to accurately estimate the amount of water available from such sources.

Food

According to Robert Feeney, many of the polar expeditions of the 1800s and early 1900s went awry because in those days people lacked a good understanding of nutrition and had yet to develop acceptable food technology.[24] Some of the problems reported by early expeditions, including exhaustion, poor judgment, and delirium, could have been produced by too few calories, too little fat, and vitamin deficiencies. The Scott expedition to the South Pole might have fared better given more generous rations, especially more fat. Arctic explorers who chose or were forced to live off the land—such as by consuming raw polar bear livers—occasionally developed trichinosis or suffered from visual and mental problems brought about by the surfeit of vitamin A from the animals' livers. Other problems stemmed from imperfect canning methods. These included spoilage in the can and contamination brought about by tin plating and the thick lead solder used to anchor the top in place. There were no simple can openers, and in the process of hacking off the tops metal fragments were mixed into the other ingredients for consumption.

Today, we know much more about nutritional requirements, food preparation, and food storage.[25] Spacefarers require good, well-balanced meals that satisfy all of their nutritional needs. They eat on the order of twenty-five hundred to three thousand calories per day, typically spread out over three or four meals. Special nutritional needs arise from the rapid depletion of calcium under weightlessness and from occasional dietary deficiencies. Inadequate potassium, for example, may have accounted for the cardiac arrhythmia experienced by Apollo 15 astronauts on the Moon. The most likely causes of poor nutrition in space are bad eating habits brought from Earth, low appetite induced by stress, and skimping on meals to reduce the number of sessions on imperfect space toilets.

Spaceflight conditions complicate the storage, preparation, and consumption of food. These activities require compact galleys and dining areas and eating utensils that succeed under conditions of microgravity.

Pots full of boiling water and pans of sizzling grease are not acceptable. Meat must be deboned and degristled, and sauces (at least those not reduced to a thick paste) are a problem—imagine the detritus that might be found on a well-used tablecloth instead floating around the cabin's interior! Food must have a long shelf life, and it must be easy to prepare—who would want to mess with dirty pots or pans? To reduce demands on the habitat's waste control system, there should be as little packaging and residue as possible. Floating strands of spaghetti or sandwich crumbs are hazardous when they clog air filters or drift into electrical circuits.

Space foods have advanced a long way since they were cut into small cubes and dipped in fat (to minimize the problem of crumbs) or stored in squeezable plastic tubes, restricting spacefarers' diet to such items as borscht, pureed meat, thin currant jelly, Tang, and "bone-bones" (now known as granola bars). Food advanced significantly from Skylab on, and today spacefarers enjoy full meals. In addition to fresh foods, the shuttle pantry includes loaves of bread and other products that have been irradiated to extend shelf life. It also holds meals that can be simply unwrapped and eaten and partially dehydrated foods that are reconstituted with a special pump before heating. NASA dietitians are aware of the importance of such factors as preference, taste, variety, and texture. Prior to flight, and in collaboration with a dietitian, shuttle astronauts plan individual meals that are then preassembled and need only be heated before they are consumed. For the ISS there have been even greater efforts to develop frozen, refrigerated, and microwavable foods that have a fresh-food taste and closely resemble their terrestrial counterparts. The meals themselves, like shuttle meals, will be served in trays with compartments for the various dishes, and eating utensils will be held on by magnets.

At some point, food will be grown in space. The menus on huge, orbiting satellites, on the Moon, or on Mars are likely to be quite different from the menus that we expect on Earth. Meals are likely to be heavy on cereals and vegetables. As we shall see, if you prefer tofu to roast beef you may be in luck.

Clothing

In addition to the pressure suits and space suits worn for protection, spacefarers wear everyday apparel suitable for shirtsleeve environments.[26] This clothing is comfortable and functional: jumpsuits, sweatshirts, T-shirts, athletic socks, and soft underwear and footwear. There are plenty of pockets to hold things, and zippers or Velcro fasteners keep pens and other items from drifting away.

Some outfits, such as the blue jumpsuits that spacefarers wear on occasion, have a uniform, quasi-military appearance, but for the most part the dress code is casual. The lack of regimentation in clothing and the availability of choice enlivens the setting and gives spacefarers opportunity to express individuality. Pins, most jewelry, and other items of adornment are out because they impose a weight penalty and could be hazardous if they broke loose and floated about. Cosmetics including hair tonic, perfumes, and colognes are undesirable, as they can add additional contaminants to the air.

Jack Stuster points out that the kinds of clothing and clothing care that we are accustomed to on Earth do not translate easily into space.[27] Clothing should be lint-free. In microgravity, floating lint that isn't inhaled wends its way to the air intake ducts, where it clogs filters and has the potential to damage sensitive machinery. Indeed, on Mir, air intake vents have been covered with an inch or two of lint, a carpet so thick that it has to be yanked off by handfuls rather than gently vacuumed away according to official procedures.

Furthermore, there is no simple way to launder items.[28] Not only is water very expensive, it won't behave right (imagine trying to run your washing machine under conditions of weightlessness), and chemical cleaning would release noxious gasses. Because of this, each change of clothes is worn for an extended period of time, and then either returned to Earth for laundering (as on the space shuttle) or sent to destruction along with other trash (as on Mir). Dirty clothes contribute to the buildup of body odor within the craft, and a greater supply of clean clothes is a common recommendation from spacefarers. It's not impossible to design laundering equipment for use in space; it's just that such equipment will be complicated and expensive. Thus far, it has been cheaper to ignore the problem of space laundries. However, it may be cheaper to launder clothes and linen aboard the ISS than to pay the cost of resupply.

Waste Management

Waste management challenges include disposing of empty packaging and other trash, and developing systems to accommodate personal hygiene needs. Accomplishing this in space requires overcoming three difficulties: the limited water supply, cramped quarters, and microgravity.

On the early flights, astronauts disposed of trash by securing it behind their contour couches, much as some of us get rid of fast-food wrappers by dumping them into the backseats of our automobiles. On Russian space stations, trash is stored in various modules that happen not to be in regular use at that particular time, or it is deposited in the Progress ferries that bring fresh supplies. The spent cargo vehicle is kicked out of orbit and incinerated on reentry. On the shuttle, trash is carefully stowed and then discarded following its return to Earth.

On lunar and Mars missions, trash can be recycled. Perhaps packing materials, containers, and the like could be designed in such a way as to have a useful life after their contents have been removed. Instructive here is Henry Ford's insistence that the batteries delivered for use in his Model T automobile be packed in wooden crates that could be disassembled and then rebuilt as frames for the car's seats.

Personal hygiene systems fall into two categories: washing facilities and space toilets. The limited availability of water and the lack of gravity make it difficult to wash in normal ways. Special showers were developed for Skylab, Salyut, and Mir, and a shower will be available on the ISS. Early showers were collapsible affairs that the bather erected around himself. Skylab's had a showerhead that essentially vacuumed up the water before it dispersed around the cabin. It was difficult to use, but it did allow the user to keep clean. Shuttle astronauts do not have a shower, but they do have a special hand-washing unit that allows them to seal their hands inside a enclosed sink.

"How do you go to the bathroom in space?" is one of the questions most frequently asked of astronauts.[29] Early astronauts wore urine collection tubes. Typically, urine was stored for subsequent chemical analysis and disposal. Such analyses made it possible to identify chemicals related to stress and to track levels of calcium and other essential minerals. Today's space suits have special tubes to collect urine; both male and female versions are available. Astronauts wear underclothes that "operate on the diaper principle," a practice eschewed by their Russian counterparts. Russians, however, are willing to take quick enemas to delay their first visit to the space john.[30]

In microgravity, bowel movements are particularly challenging. For Gemini and Apollo flights, astronauts taped plastic bags to their rears. When they were done, they had to break open a small capsule within the bag, and, through massaging the contents from the outside, work the substances into the feces. This halted natural processes of microbial growth, and kept the bags from exploding. All of this had to be done in quarters that were not that much larger than the cockpit of a sports car, constantly rubbing elbows with a fellow astronaut. To this day, Soyuz lacks full toilet facilities, so passengers must use plastic bags during the two days or so that it takes to catch up with a space station. Plastic bags, something like those used on airplanes, are used for collecting vomit.

Skylab marked the introduction of space toilets, and these served as prototypes for the models on the shuttle today. In lieu of flushing with water, waste is forced "downward" by means of air pressure. On Mir, spacefarers urinate into a funnel-shaped device, but eliminating fecal matter requires careful positioning over a two-inch hole and assuring a good seal. Thus far, spacefarers have been forced to relieve themselves in semipublic or to assume uncomfortable positions in tiny cubicles. On the Moon, and on Mars, there is at least some gravity, so we will not be quite so reliant on exotic techniques to draw wash water and bodily wastes downward.

In Situ Resource Utilization

If we want to establish ourselves on the Moon or Mars, we have to learn how to live off the land. This will depend on in situ resources utilization (ISRU). We have already seen how it might be possible to liberate oxygen or water to help support human life on the Moon or Mars. Most of the processes required for ISRU are patterned after industrial processes in use on Earth.

Thomas Meyer and Christopher P. McKay describe many justifications for ISRU on Mars and add that the sooner we begin exploiting Martian resources the better.[31] The most obvious benefit is that ISRU saves transportation costs. If, for example, we can use Martian resources to develop rocket fuel for the trip back to Earth, we can save the cost of transporting fuel from Earth to Mars for the trip home. ISRU also means that recycling equipment will not have to be quite so efficient and that habitats will not have to be so tightly sealed. Because the Martian atmosphere could help us replenish oxygen and water, we will not be quite so concerned if some of it leaks out of the habitat, is wasted, or is otherwise lost. Rationing need not be as strict as would be required if we were completely at the mercy of imported supplies, and the settlers will have a deeper sense of self-reliance. With ISRU, life and death will not hinge on the arrival of the next supply ship.

ISRU depends on the availability of power and raw materials. As far as we can tell, life never took hold on Mars: there are no oil or coal deposits or other fossil fuels. Although Mars is farther from the Sun than is Earth, its thin atmosphere means that, except during dust storms, more of the light that does arrive can get through to solar cells on its surface. Additionally, Mars is a windy planet, so carefully designed windmills could generate power. Chemical fuels can be derived from the Martian atmosphere and used for local transportation as well as for the return to Earth.

Because of the way Mars was formed, we do not expect it to be rich in metals other than iron. In the future, Martians may be able to get metals by mining the asteroids and, because of Mars's low gravity, they should have a relatively easy time of getting to the asteroids to do this. However, we can accomplish much with the materials already there. Iron production could begin early on, with production of high-grade steels to follow. Martian soils will yield the materials for low-grade ceramics and general purpose glass. The composition of Martian soil is such that it can be used as cement and even formed into a concrete-like substance called duricrete. Using raw materials like those available on Mars, duricrete has been formed into building blocks on Earth. When reinforced with fibers, it is both durable and strong.[32]

Over time, agricultural production will become possible. Martian soil will be capable of growing plants after conditioning with fertilizer derived from local resources, including wastes from the habitat.[33] Compression of the Martian air, which is predominately carbon dioxide, yields a greenhouse-type atmosphere congenial to plants. Agricultural researchers have already made great strides in genetic engineering, and perhaps their talents will help us develop crops such as the "Martian potato."[34] This plant's hearty roots could reach down beneath the frozen surface to liquid water, and its huge leaves could soak in the Martian midday sun.

Biospheres

Different activities—such as human habitation and plant growth—can be combined so that each benefits the other. A simple example is connection of a habitat to a greenhouse: air exhaled by humans in the habitat provides carbon dioxide, which benefits the plants; the plants generate oxygen, which benefits the humans. The two forms of life—animal and plant—are synergistic, in that they do better together than either one would do alone. The development of large, synergistic systems is important for ISRU.

Biosphere 1 (also known as Earth) is a large, self-regulating system containing millions of life-forms. Biosphere 2 is a 3.15-acre habitat constructed in the Arizona desert.[35] It too is self-regulating and can sustain people in comfort for extended periods of time. Biosphere 2 is a closed system that includes living quarters, a breathable atmosphere, plenty of water, diverse mini-ecologies (savannas, plains, streams), and the means for agricultural production. Natural processes are intended to keep the air and water fresh and put healthy food on the table.

In September 1991, an eight-person crew began a two-year stay in Biosphere 2. Air pressure was maintained by a patented system involving two artificial "lungs." Air was recycled and purified by means of "soil bed reactors" that blew air up through the dirt so that microbes could scrub out toxins and maintain air quality. Water, including that from human waste, was recycled through marshes, where microbes and aquatic plants purified it. Biosphere 2 even has a "nerve center"—a computer programmed to monitor and analyze atmospheric conditions and water quality and then put servomechanisms, pumps, and valves to work to keep everything in good order.

One goal was to produce a varied, nutritionally sound diet using sustainable agriculture. The crops selected would not exhaust the soil and make it difficult or impossible to grow more crops in the future. Pesticides, herbicides, and other toxic substances were replaced by non-poisonous sprays and by insects that would ignore the crops but prey upon the crops' predators. Biractoreans grew many different grains, fruits, and vegetables, including sorghum for animal fodder, sweet potatoes, peanuts, beans, soy, squash, bananas, and herbs. Goats were available for meat, milk, and cheese, and chickens for meat and eggs. Fish were raised too.

As the mission progressed, some of the ecosystems changed over time, and a mechanical device had to be brought in to help manage fluctuations in carbon dioxide. At one point, additional oxygen had to be pumped into Biosphere. Each crewmember had to work about 66 hours a week—as compared to 56 hours on the shuttle—and two-thirds of this was devoted to basic maintenance activities such as growing food, leaving less time to do research than had been anticipated. Although Biosphere 2 did not remain sealed for the full two years, it gave us new insights into self-contained bioregenerative systems.

Planetary Engineering

The rule of thumb for biospheres, states Martyn Fogg, is "the bigger the better."[36] In comparison to small biospheres, larger biospheres accommodate more people and grow more crops. A larger biosphere is less responsive to imbalances and perturbations; for example, air loss at a rate that could doom a tiny biosphere might pose little or no risk for a huge one. Furthermore, large biospheres should be more enjoyable. An enclosed environment where it takes a day or so to walk from side to side is more fun than one that can be traversed in just a few minutes. Given the motto "the bigger the better," our ultimate goal should be turning an entire moon or planet into a shirtsleeve environment. Such a planet would not have to imitate Earth, but it should have an atmosphere that allows us to breathe without special apparatus, a temperature that permits us to wear normal clothing, and the materials for sustaining high-grade agriculture.

Planetary engineering refers to activities aimed at planetary change: adjusting temperatures, changing atmospheres, and creating conditions that alter the ecology of a planet.[37] *Terraforming* refers to making planets or moons more Earth-like, that is, more welcoming to humans. While either is ambitious, notes Fogg, we have already transformed our own planet. We have changed Earth's landscape through enormous pit mines and through agriculture; we have rerouted waterways through systems of dams, locks, and canals; and we have released tons of hydrocarbons and other chemicals into the atmosphere, creating global warming and cutting holes into the ozone layer. Earth today is very different from Earth of a few hundred years ago. The major changes that we have made on our home planet were unplanned and unintentional, but the changes that we hope to make on other planets would be thoughtful and deliberate.

Marshall Savage's comment that terraforming Mars will "be like reconstituting freeze-dried soup, we will just add water" may be a bit of a simplification, but he is right that Mars invites terraforming.[38] To make Mars more Earth-like we would raise the temperature on the surface, increase the availability of water, and promote a thick, oxygen-rich atmosphere.

One approach is to make Mars hospitable for microbial life. Like the blue-green algae that made Earth habitable for humans, the microbial life placed on Mars could help develop a usable atmosphere, although if it were left to do this by itself it could take thousands of years. *Ecopoiesis* refers to the purposeful production of an ecosystem or biosphere, whether or not it is suitable for humans.[39] This raises interesting questions about environmental ethics. Is it acceptable to "play God" and seed life on barren planets? What about rearranging the ecologies of planets that already have life? D. MacNiven points out that from the "cosmocentric" perspective this is a tough call that requires balancing the preservation and enhancement of human life against the intrinsic value of another planet and its ecology.[40]

Mars can be sprinkled with uniform gray powders that absorb heat and cause permafrost or ice to melt. Other strategies include using microwaves to melt the ice or detonating war surplus nuclear warheads to kick up dust and raise the temperature.[41] Long, thin tubular explosives inserted lengthwise into the ground and then detonated would melt deep permafrost much more quickly than procedures aimed at increasing temperature on the red planet's surface. Over time, as the planet warms up, algae, and later on forests, would reduce carbon dioxide and increase oxygen.

Light and warmth could be increased by orbital mirrors that would reflect additional sunlight on the Mars's surface. (Indeed, on two occasions cosmonauts aboard Mir experimented with prototypical mirrors in the hopes that one day this technology could be used to liberate resources frozen in the vast expanses of Siberia.)[42] Here, Fogg favors orbiting a number of relatively small mirrors 250 kilometers in diameter but almost unimaginably thin, rather than launching a huge new "sun." By the time such mirrors could be fabricated and installed, we may have already gained considerable experience in space mining and manufacturing, so this may not be as impossible as it seems. Mars will not lose its identity after terraforming, concludes Fogg: "Mars will still be Mars, an exotic and alien world providing a new and unique stage for the dramas of life and civilization."[43]

CONCLUSION

Despite the lethality and improvidence of outer space, most contemporary space habitats offer occupants livable environments. So far, life support systems have been at least adequate given mission requirements. Spacefarers have wished for better toilet facilities on delayed transit vehicles, worked under freezing and sweltering conditions, and gone on short rations. Yet, apart from the crewmembers of the first Salyut, who were asphyxiated on Soyuz's return, there have been no losses of life due to the failure of a life support system in flight.

Clearly, life support systems have improved over the past few decades. Perhaps most notable are improvements in diet and in personal hygiene and waste management systems. Unlike the early spacefarers, today's spacefarers have a varied menu that includes warm meals, as well as the opportunity to shower aboard space stations, and toilets that are at least a rough approximation of the kind they use on Earth.

Provisioning space missions is always expensive, and recycling cuts costs. On future missions to the Moon and Mars, it will be essential to capitalize on local resources. This will dramatically reduce transportation costs, relax design requirements, act as a safety net against disruptions along the long supply routes from Earth, and confer a sense of independence.

Following the lead of Biosphere 2, we should find ways to make different forms of life prosper as a result of each other's presence. Biosphere 2, which uses natural processes for purifying air and water and enables varied agricultural production, may contain the basic ideas for establishing ourselves on Mars.

In the very distant future, we may be able to transform entire planets. Although terraforming originated in science fiction, it has gained the attention of a small group of scientists. Later on we may be able to create very large new biospheres by reengineering planets and moons to make them fully hospitable to human life. How realistic is it to expect terraforming? According to Haym Benroya, terraforming is as much a philosophy as a science.[44] Many of those who favor terraforming have strong backgrounds in physics and biology but become somewhat vague when it comes to engineering, leaving the specifics to future generations. Benroya's own assessment as an engineer is that many of the technologies we would need exist in incipient form and can be expected to mature and become workable over the next few centuries. *"What can be imagined,"* he writes, *"can be built."*

CHAPTER 5

HABITABILITY

After his selection for a Gemini mission, Michael Collins received, as a joke gift, a garbage can with two tiny Gemini-style windows painted on its sides.[1] Although spacecraft have improved substantially since those days, they remain relatively small affairs that lack many of the comforts and conveniences of everyday life. Thus, balanced against the joy and excitement of flight are the limitations of spacecraft living. These limitations fall into three general areas.[2]

First, spacecraft offer low levels of environmental comfort. This stems from cramped quarters, poor ventilation, temperature and humidity imbalances, high ambient noise levels, and makeshift or do-it-yourself services that substitute for the professional services available on Earth. Although spacefarers can make some choices and do have some variety, life in space can have something of an institutional flavor, which includes shared sleeping quarters, limited wardrobe, regulation supplies, and regimented activities.

Second, spacefarers are, to one degree or another, inconvenienced. Examples include rationed showers, infrequent clothing changes, restricted diet, and in the past, unappetizing meals served at room temperature. Because facilities are limited, spacefarers may have to wait their turn to communicate with home or peer out of a window. Forms of emotional release available to people on Earth (alcohol, tobacco, drugs, sexual contact, a wide range of recreational opportunities) are available on a limited basis at best.

Third, spacefarers may have less of a sense of personal control over their environments and their lives than they normally experience on

Earth. They have restricted access to news and limited opportunity to converse with family and friends. Apart from demanding work assignments, there is little or no opportunity to "go outside," and the usual weight and storage considerations limit the personal items that they can bring along. Many activities are assigned, and the work schedules are often heavy. Their choices are constrained by a plethora of rules and regulations.

Habitability is a general term that connotes a level of environmental acceptability. An environment is habitable to the extent that it accommodates its occupants and is acceptable to them.[3] Habitability rests in part, but not exclusively, on the objective characteristics of the setting. Of course habitability depends on protection against radiation and physiological deconditioning, the reliability and effectiveness of the life support systems, and other factors described in the past two chapters. It is affected also by the design and layout of the habitat, aesthetics, illumination, noise and odor control, equipment and supplies, and many other considerations.

There are powerful reasons for seeking a balance between habitability and cost instead of always opting for the lowest cost. One is the increased demands associated with long-duration missions. Compared to people who visit space for a few days or weeks, those who live in space for a month or more need better accommodations. Over the long run, low levels of habitability wear people down. The Russians found that by improving food, waste management, and entertainment systems, and by trying to create a more "homelike" interior on later Salyut space stations, they reduced boredom, irritability, and some of the other problems associated with earlier flights.[4]

Second, primitive conditions acceptable to explorers are not likely to be acceptable to the scientists, laborers, and other people we will want to attract for future missions. This basic idea was captured by the Polar Research Board in a discussion of life in polar regions. Initially, the goal is to ensure human survival, but when the first explorers are replaced by settlers we shift emphasis to "providing a situation in which a person can expect to be healthy, happy, and effective in family life, work, and community relationships, without crippling emotional symptoms such as fear, anger, loneliness, envy or greed."[5] Primitive conditions would be even less acceptable for space tourists, who will demand and expect a good time.

Habitability depends also on the characteristics of the spacecraft's occupants: their personality, attitudes, and other psychological factors that shape their perceptions of environments. Thus, the same place that one person finds harsh and threatening, another person might find interesting or challenging. Different people, as Peter Suedfeld points out, experience the same environment in different ways.[6] Their experiences of environments are more important for understanding their mood and behavior than are the objective qualities of the environments themselves.

Habitability is related also to the circumstances of the occupancy. To estimate habitability we need to know whether or not "being there" is a common or unusual occurrence, the duration of the occupancy, and the goals served by staying there (for example, scientific advancement or national prestige). A place may seem more habitable when one is engaged in an activity that is important and exciting.

Putting this all together, habitability depends on knowing more than "What is the environment?" It also depends on knowing "Who are the environment's occupants?" and "What are the situations and circumstances surrounding the environment's use?" This means that improving objective living conditions is only one way to promote favorable attitudes toward life in space. Alternatively, we can seek people who find the environment's limitations easier to bear, or change mission configurations to make the living conditions seem more acceptable. We know, for example, that college research participants view the same harsh spaceflight conditions as more acceptable for brief as compared to extended missions and for small as compared to large crews.[7]

At first, all we needed to do was get people into space and then bring them back alive. Now, with people staying in space for months at a time, and with the prospects of space tourism and space settlement, we must consider how to satisfy a wider variety of human needs. Drawing on Clayton P. Alderfer's work, our task is to accommodate spacefarers' existence, relatedness, and growth needs.[8]

Existence needs are basic human requirements in order to stay alive and healthy and to feel secure. In effect, most of my earlier discussion of hazards and countermeasures and of life support systems relates to accommodating existence needs. To these ends engineers must protect spacefarers against the elements and give them reliable, effective life support systems. Existence needs have been fulfilled in the past, but now we seek to satisfy them on a longer term basis, by greater margins, and with higher quality results.[9]

Relatedness needs are people's requirements in order to form rewarding, positive and enduring relationships with other people and to enjoy their affection and respect. They are satisfied by developing acquaintances and friendships and gaining full membership in effective teams. Still, people sometimes need solitude. To satisfy relatedness needs, then, crewmembers must be able to regulate social interaction. Thus human factors experts try to counteract the effects of isolation from family and friends, build good will within the crew, and make allowance for privacy.

Growth needs are people's requirements for variety, challenge, self-expression, and personal development. Satisfying these needs helps people feel good about themselves. To accommodate these needs, human factors experts encourage individuality and "humanize" spaceflight: they provide the voyagers with comfortable, aesthetically pleasing environments that offer opportunities for education and self-development. Although it may seem that no activity in space lacks glamour and excitement, this may not be true for people who undertake multiple missions or who stay in space for many months at a time.

ARCHITECTURAL CONSIDERATIONS

Architecture refers to the design and construction of buildings and other structures, including spacecraft and habitats. Architecture determines form, function, and aesthetics. Architectural considerations related to habitability include the size and configuration of the habitat, the volume and layout of the work and living areas, and design features such as lines, colors, and the availability of windows or viewports. Each of these plays a powerful role in perception of environmental acceptability.

Wolfgang Preiser's environmental design cybernetics is a general approach to design that has high applicability to space.[10] Preiser views design as an "ongoing experiment," a cyclical process involving planning, programming, design, construction, occupancy, and postoccupancy evaluation used as feedback to initiate a new cycle. The implication of environmental design cybernetics is that to "humanize" the space environment we must seek user input and foster designs that can be modified on a trial-and-error basis. Designs should reflect the priorities of the users, not the priorities of people who will never live in the spacecraft. To reckon with individual and cultural differences, we avoid the tendency to homogenize and instead favor varied or differentiated environments that allow latitude for further refinement by the users.

Forms and Configurations

Space habitats are launched into space and are either self-contained or joined together, like Tinkertoys or Legos, to form larger configurations. The basic form is the cylinder. The sense of living aboard a railroad train is reduced by the addition of modules that jut outward from the center, as in the case of Mir, or by interconnected cylinders and spheres. We can expect many different configurations at lunar or Martian bases, as a variety of habitats are first deployed and later built.

Habitats must have adequate internal volume to support work, living, and recreational activities. Estimates of the amount of space required per person vary, in part as a function of mission duration. According to one NASA estimate, required habitable volume per person increases as a function of mission duration, but peaks and levels off at approximately six months.[11] (Habitable volume refers to usable space, not space occupied by machinery and stored equipment.) According to these estimates, 5 cubic meters per person is tolerable and 17 cubic meters is optimal for a six-month mission. Thus, a minimal level for a 6-person crew would be 30 cubic meters and an optimal level would be 102 cubic meters. More recent estimates suggest that as we further increase the crew size and mission duration, space allocation should rise rapidly: a ten-person crew on a yearlong mission would require about 68 cubic meters per person, or 680 cubic meters for the entire crew.[12]

Planning begins with a comprehensive activity audit to identify all of the living, working, and recreational activities that will occur throughout the course of the mission. This is followed by determining activity envelopes; that is, the amount and configuration of space required for each activity. In all locations, equipment must be accessible and easy to operate. Work stations should have enough clearance so that workers can use keyboards, handle tools, and apply whatever force is necessary to get the job done. Extra space is required where two or more people are expected to work together.

Designated areas should be large enough to accommodate social needs. People try to regulate their physical distance from one another to achieve maximum comfort. The distance that is the most comfortable depends on such variables as cultural background, the relationship of the people to one another, and the type of interaction: formal, informal, or intimate.[13] A comfortable informal conversation requires at least one meter of separation of the area's occupants. However, comfortable distances vary as a function of culture, and since most future missions will be international, these variations need to be taken into account. In comparison to Americans, representatives of British and northern European cultures require greater separation, while members of Mediterranean, Middle Eastern, and Asian cultures seem to require less distance from one another.[14]

Simulations involving college students suggest that microgravity affects the interpersonal distances that people find the most comfortable.[15] In microgravity, people can float into almost any position, so you could end up conversing with someone who appears to be upside down. Participants in the study were shown dolls that were standing in heads-up positions common on Earth: face to face, side by side, or at a right angle with their feet on the floor. They were also shown dolls that were placed in unusual orientations made possible by microgravity; for example, both people "upside down" (relative to the space station's "floor") or one person "right side up" and the other "upside down." Research subjects judged that greater interpersonal distance was required for holding a conversation when the two participants were oriented to one another in the unusual ways made possible by microgravity.

Careful planning is required in situating task-dedicated areas. There should be good separation between sleeping areas and working areas. In this way, spacefarers who are trying to rest will not be disturbed by those who are trying to work. Similarly, public areas such as wardrooms or galleys should be separated from the sleeping quarters, again to protect those who are trying to rest. Personal hygiene facilities should be outside of high-traffic areas, especially those used for cooking and eating. After a new toilet was installed on Mir, cosmonauts and their guests quit using the initial toilet, which was only two feet from the table used for meals.[16]

Because interior space is limited, some interior areas must be multipurpose. These areas should be made definable and redefinable by their occupants. Flexibility is enhanced by carefully planning "hard" architectural features (interior dimensions, walls, hatches, etc.) and the use of lightweight or "soft" features (such as screens, moveable partitions, fold-down and pop-out furniture) to modify the interior. Occupants can use personal items to "stake out" areas in the same way that students use textbooks and other personal items to gain temporary control over large portions of a library table. (However, unlike in a university library, these items may have to be held down by Velcro.)

Deployable Structures

Structures at lunar, Mars, and other planetary bases are likely to undergo three generations of evolution. Initially, the spacecraft will serve as the habitat. Just as a tourist might pull a Winnebago to the side of the road to camp for a few weeks, spacefarers will in effect park a spaceship or lander on the Moon or Mars to serve as their initial home.

The second generation (which may appear close on the heels of the first generation) is likely to consist of special structures brought from Earth to increase living and working space. This is comparable to erecting tents and prefabricated buildings on Earth. Although it might be necessary to move rocks, scrape the regolith, or dig a hole or two, the structures themselves will be imported.

The third generation will consist of quasi-permanent or permanent structures constructed of local materials. This construction phase would be analogous to felling trees and erecting cabins on Earth. Over time, the functional equivalents of huts or cabins would be replaced by bigger and better structures arranged into towns and cities.

Deployable structures are the key to the second generation.[17] These must be lightweight and compact for transit and then, like a pop-up tent, easy for a small crew to erect. One option is inflatable structures that are simply placed on a level spot on the ground and filled with bottled air or, if a local atmosphere is available, by means of a compressor. Such structures proposed for human habitation, greenhouses, and storage huts could be hemispherical (like igloos) or semicylindrical (like Quonset huts). They could be anchored by a deep, even layer of regolith on the built-in fabric floor, which itself would be covered by additional flooring material (something like a rug to walk on). An advantage is that walls or partitions and even furniture (such as sleeping alcoves or benches) could be built in. Many such units could be joined together, forming a small town. Inhabitants could protect themselves from radiation by burying the living areas underground. Because ultraviolet radiation and corrosive dust are likely to cause inflatable habitats to degrade over time, people may choose hard structures for habitation and use inflatable structures for storage and work areas.[18]

Willy Z. Sadeh and Marvin E. Criswell have addressed some of the basic design requirements for inflatable structures on the Moon. These would consist of thin, inflatable membranes containing thicker inflatable frameworks. These modules could be joined together to produce structures of almost any size or shape.[19]

Gary Brierly and his colleagues have designed a deployable habitat that would be sent on ahead to Mars to await the spacefarers' arrival.[20] This would consist of a central core that, in its compact state, would be an octagon approximately 27 meters high and 2 meters wide. It would be shipped containing supplies for use at the destination, automated systems to self-deploy, and television cameras that would permit observers on Earth to monitor the autoconstruction process. During transit, the core would be filled with pressurized gas, which would be released as breathable air when the structure is occupied.

On arrival, automated excavation machinery would dig holes to anchor the habitat and pile regolith on the habitat's roof, forcing it downward and outward into a squat habitat up to 16 meters in diameter. A tubelike metal core would serve as the air lock. Periscope-like windows would allow light to enter via a series of mirrors while blocking radiation. All that the incoming spacefarers would have to do is check and make sure that there were no problems that failed to show up on the TV monitor. If everything worked, they could move right in.

PRIVACY

Invariably, privacy is an issue when people are cooped up in small, cramped quarters. Privacy exists to the extent that an area's occupants can limit unwanted forms of social intrusion. Privacy serves multiple functions.[21] First, although work often requires the concerted action of groups of people, there are occasions when people work best alone. Privacy is helpful for achieving concentration or focus when other people are distracting. Second, as revealed by physiological measures, other people tend to be exciting in the sense that they increase physiological activation or arousal. Getting away from other people serves to reduce this excitement. Privacy promotes "rest and recuperation." Third, privacy helps us control the images that we project to one another and thereby regulate the relationships we have with them. Getting away from other people decreases the chances that behaviors such as anger or weeping will make us look bad. Sometimes we retreat from one another to "get offstage" so that we don't have to monitor ourselves quite so closely or be quite so concerned about what other people will think of us. Finally, groups, like individuals, need privacy. People may wish to interact with one another without the rest of the crew watching. Simple examples include holding confidential or intimate discussions, venting emotions, and providing critical performance feedback.

Even as there are conditions under which we seek to retreat from others, there are times when we welcome their presence. Sometimes we seek out other people precisely because they are a welcome form of diversion. Sometimes we seek them because we want useful information or emotional reassurance. Thus, even as we consider people's need for privacy we must also consider their need to be with and interact with other people. Most discussions of extended duration missions tend to recommend private quarters for each spacefarer. There, people can sleep, dress or change clothes, read, watch TV on a small screen, and use a private computer workstation. As much as possible, individual occupants should have control over heat, ventilation, and lighting. There should be adequate storage space for personal possessions, and once inside the quarters, the occupant should be unseen and unheard by people outside.[22]

On early craft, spacefarers tried to sleep cheek-by-jowl in their neatly aligned lounge chairs. On the shuttle there is a little more room, and the astronauts place their sleeping bags throughout the cabin. Skylab had three tiny adjacent individual sleeping quarters, enhancing the sense of privacy. However, the fabric doors were fastened by Velcro, and the ripping sound made as people entered and exited disturbed other crewmembers.[23] Mir has sleeping quarters behind pull-curtains, but not enough for each spacefarer. Guests sometimes find solitude by berthing down in one of the attached modules. Plans for the ISS call for individual sleeping quarters a little larger than telephone booths.

At the other end of the continuum are wardrooms and other public areas where all crewmembers can gather. These are useful for meetings to discuss issues of interest to the entire crew and for ceremonial occasions. Here seating arrangements should allow people to cluster together in subgroups, and to arrange themselves vis-à-vis each other in such a way that they can avoid or enter the fields of view of the others. We also need areas where small groups can meet apart from the rest of the crew. One alternative is to use tacked-on modules as semiprivate meeting places. For example, two or three spacefarers might retreat to a laboratory module during periods when no experiments are scheduled.

Not everyone requires the same degree of privacy. Some people are able to tolerate high-density living conditions or they have psychological mechanisms for shutting others out. Douglas Raybeck points out that in some Middle Eastern cultures, people prefer to live in one large room.[24] If they wish to withdraw, they simply quit talking. Japanese and members of other Asian cultures tend to thrive despite living in high-density cities: focusing inward through meditation and other processes helps them avoid a sense of crowding. Among North Americans, all-male groups tend to require more space and more privacy than do all-female or mixed groups. The same packed rooms that men describe as crowded women describe as cozy. Nonetheless, Raybeck points out, members of all societies must have some means of achieving solitude. We will have to make architectural allowances for privacy.

Surveillance of the crew by outside monitors is another issue. Ground-based managers may consider it useful to monitor onboard activities to make sure all is well, to find ways to improve future missions, and to intervene in an emergency. On the other hand, such surveillance may constitute an unwelcome intrusion. Charles Berry, an early NASA physician, recommended that mission control keep only one section of the spacecraft under surveillance, so that astronauts could escape to private areas.[25] Other possibilities include making it possible for spacefarers to shut off video cameras and microphones, or introducing two-way surveillance systems to reduce the stresses associated with being seen by outsiders without being able to see the outsiders in return. Tourists and settlers, however, will find surveillance of private quarters unacceptable.

FUNCTIONAL AESTHETICS

An aesthetic design—one that is beautiful, in good taste, artistic—may seem unnecessary for a highly motivated, task-oriented group such as spacefarers. However, as Yvonne Clearwater and her associates point out, aesthetics is not merely a matter of satisfying personal tastes. *Functional aesthetics* refers to design elements that counter some of the less positive aspects of living in space and that are intended to boost spacefarers' performance and morale.[26]

Microgravity enables design configurations that are not possible under normal gravity. Even though to some spacefarers "up is where my head is," configurations that clash with those that we are used to on Earth can be confusing. Interior layouts should constitute coherent frames of reference so that occupants can immediately orient themselves within the spacecraft. This is particularly important under emergency conditions when people need immediate, accurate bearings so that they can take prompt and effective action. We must be wary of topsy-turvy layouts that seem interesting or efficient but are disorienting in use. *Environmental legibility* refers to the extent to which a design imparts a sense of coherence and stability.

High legibility can be achieved by incorporating cues that impart the sense of a true vertical despite microgravity. To some extent, this can be done by arranging the environment as if it were in a gravity field. Christopher Barbour and Richard Coss found that it was possible to establish a vertical by such simple expedients as painting the "floor" and the lower half of the wall a darker color than the "ceiling" and the upper half of the wall.[27] This practice, which they tested with an eye to the ISS, was used also in Russian space stations. The basic module in Mir, for example, has a three-tone paint job, with the darkest shade of green on the floor, an intermediate shade on the walls, and the lightest shade on the ceiling.

Certain design elements can increase the perceived spaciousness of an area and thereby help offset some of the problems associated with tiny quarters.[28] Unfortunately, given the tubular configurations of most spacecraft, rooms with curved walls seem smaller than rooms with angular floor plans. Irregularly shaped rooms seem to have greater volume than regularly shaped rooms with the same square footage. Light, pale, or unsaturated colors create a greater sense of spaciousness than do dark colors, and in comparison to a vertical configuration (with rooms stacked on top of each other, something like a European townhouse) a horizontal configuration (reminiscent of a single-story North American home) accentuates interior space.

Studies of people in a high-fidelity space station mock-up and Australians wintering over in Antarctica suggest that art helps people adapt to unchanging, monotonous environments.[29] These studies found that even dark-colored pictures were not degraded by the low levels of illumination expected on some parts of the ISS, and that people enjoyed pictures that were placed at unusual angles. This suggests that floating around in microgravity will not undercut people's ability to enjoy artwork. Indeed, exploring artwork from many novel angles may sustain interest.

Salyut 6 cosmonauts had access to a good library of videotapes, but found those depicting natural scenery especially appealing. The Clearwater and Coss studies also found that viewers preferred landscapes that gave a sense of spaciousness. Photographs were more effective than paintings, and both dry and glittery landscapes were received more positively than themes involving humans and animals. Depth of field was important: people liked landscapes and other pictures that gave an impression of great depth irrespective of the picture's subject. This confirms an earlier finding that landscapes that gave an impression of depth were popular on nuclear submarines.[30]

Artworks can serve the useful function of reinforcing spacefaring values. Because their huge size and heavy cargoes tend to keep them from bobbing around in the water, a supertanker may not seem all that shiplike to the people onboard. To remind sailors that they are at sea, some ships' walls are adorned with pictures of other ships, such as Spanish galleons, that have greater nautical ambiance. Similarly, navy ships carry pictures that have maritime themes and reinforce navy values. The aircraft carrier USS *Carl Vinson,* for example, has some gratuitous wood and a few highly polished brass accents; gray steel and plastic are more functional, but less evocative of the navy's long and proud tradition. Submariners are surrounded by decorative items that reinforce the value of the submarine service. Similarly, artwork with spacefaring themes may bolster the morale of spacefarers. Perhaps some future crew will include an artist in residence.

Paul Klaus proposes introducing microenvironments into barren, isolated settings.[31] These are compact living representations of landscapes, including miniature mountains, waterfalls, streams, and grasses. Unlike two-dimensional pictures of the outdoors, microenvironments offer a range of sensory experiences: true depth, touch, and even smell and taste. Previous research, he notes, shows that people who are inside buildings do better if they can grow something, and it will be possible to cultivate a microenvironment on the Moon or on Mars long before agricultural greenhouses come on-line. Tending the microenvironment is a good hobby (cosmonauts enjoyed growing things in Russian space stations), and there could be side benefits. Flowering plants could mask unpleasant odors, and the crop could include herbs for food seasoning.

Windows or viewports are very important. Richard F. Haines finds that windows help people satisfy their curiosity and become oriented to the external environment.[32] They make it easier to complete actions (such as docking) that would otherwise be very difficult. By admitting external light and making it possible to look off in the distance, windows open up interiors and make them seem less confining. Windows help preserve a sense of connection with Earth. They reduce mystery and alleviate fear of the unknown, and they promote peak (transcendent) experiences also known as overview effects (chapter 1). Looking out the window makes it possible to avoid intense social interaction and, to some extent, tune out other people. But because the holes for windows are points of structural weakness (two of the world's first passenger jets, DeHavilland Comets, disintegrated because of a concentration of stress at the corners of the windows), virtual windows offer a promising compromise. These might be large, flat, LCD screens that display pictures transmitted from a TV camera mounted on the exterior wall. They could double as display units for hundreds of thousands of artworks stored on CD-ROM, or for live televised action from Earth.

Abundant evidence shows that isolated people enjoy looking out of windows. Aquanauts in underwater habitats enjoyed peering out of portholes, and as the former navy psychologist Benjamin B. Weybrew reports, just a few moments of looking out of the periscope of a submerged submarine boosts crew morale.[33] Astronauts are no exception. During the Mercury era, engineers were afraid that the windows might crack, but the astronauts fought vigorously and successfully for them.

Viewing Earth was a favorite off-duty activity aboard Skylab. Salyut 6 cosmonauts alleviated boredom by spending many off-duty hours at station windows, glimpsing spectacular views of Earth and the aurora borealis. Similarly, cosmonauts Berezovoy and Lebedev preferred looking outside to watching video movies and telecasts of artistic performance.[34] It is not clear, though, that windows will offer the same level of benefit to people who are so far away in space that Earth is not readily identifiable and the view outside does not change appreciably day after day.

Lighting

Engineers must overcome two hurdles in order to provide spacefarers with ample lighting. First, until we can do a better job of tapping solar power and other sources of free energy, electricity will be somewhat limited. During his Mercury flight, John Glenn had tiny lights sewn into the fingers of his space suit gloves so he could see the instrument panel on the dark side of Earth. Some areas in Skylab were so poorly lit that astronauts had to use flashlights to see well enough to get their work done. Second, electric lights radiate heat into an environment that is already stifling and, in this way, add to one of the big challenges of life support.

NASA guidelines call for levels of illumination that afford good contrast but minimum glare.[35] In work areas, light should be bright and uniform. Wide spectrum, or "daylight light," is efficient for areas requiring bright light, while "warm white light" works well in areas that call for low light.[36] People associate lower levels of illumination and redder, or "warmer," sources of light with nighttime and social activities, so warm white light might be useful in living and off-duty areas.[37] Additionally, illumination levels within two adjacent areas should not be so disparate that workers moving between them have difficulty adjusting.[38]

Skylab had over eighty fixtures that used fluorescent bulbs.[39] Compared to incandescent lights, fluorescent lights are efficient and cool. Fluorescent lights are less sensitive to vibration and have a longer life. They allow more accurate color perception, and can be used to mimic terrestrial lighting conditions. However, incandescent bulbs work better in very cold temperatures, so they are used in air locks and outside the vehicle, where they give light in the shade or at night.

Area lighting helps regulate social activity. It can break up large areas into smaller public and semipublic areas. Variations in illumination demarcate an area or "set off" an individual or small group from other people in the same or adjacent areas. For example, a spacefarer who is able to reduce light in a given area can, in effect, distance himself or herself from other crewmembers. (Did you ever shut off your reading light in a passenger plane as a signal that you did not want to talk to the person in the adjacent seat?) User-controlled lighting is in the interests of privacy, again with the restriction that the lighting contrasts in public areas should not be so great as to make it difficult to move from one place to another.

Sound Control

There are two basic sound-control problems in space.[40] One is protecting hearing during takeoff, when rocket blasts (which sound like muffled explosions within the ship) can reach 120–130 decibels, substantially louder than music at a rock concert. Although most of the sound is *behind* the passengers and effectively disappears when the spaceship reaches supersonic speeds, there is still the sound of explosive charges (sounding, according to William Pogue, like train wrecks) separating the spent boosters in multistage rockets, loud noises from the pumps, the swiveling of the engines, and the creaking of the air frame.[41]

Even after the engines shut off there tends to be a high level of ambient sound within spacecraft, on the order of 70–90 decibels. This is due to the expansion and contraction of the structure as a result of temperature fluctuations, and the whirring of pumps and fans and other life support equipment. Ambient noise is more of a problem in compact spacecraft that maintain full atmospheric pressure, such as the shuttle, than in craft with reduced atmospheric pressure. In large habitats with low air pressure, all sounds taper off rapidly as one moves away from their source. This means that a nearby companion's voice may sound faint and far away.

More than half of the thirty-three shuttle astronauts in Willshire and Leatherwood's survey indicated that shuttle sound levels interfered with speech, disturbed sleep, and were a source of annoyance.[42] Seventy-four percent of the respondents recommended lower background noise on the shuttle, and 93 percent felt that it should be quieter on the ISS. Guidelines for the sleeping areas in the ISS recommend sound levels below 35 decibels if possible, and sound levels below 55 decibels in laboratory modules.[43] Unfortunately, the ISS is much noisier than this, and shortly after the first modules were launched a work crew arrived with duct tape, padding, baffles, and other materials to try to reduce the noise to an acceptable level.

Speech has to stand out against background noise, so we must assure that speakers can be understood. This is particularly important for international crews who may have difficulty understanding one another under the best of conditions, never mind when their speech is masked by background noise. Clarity is important also if we wish to use synthesized speech, which tends to sound flat and unemotional and lacks backup in the form of nonverbal cues.

Two of the most common methods for reducing noise or vibration (for example, minimizing the amount transmitted between two adjacent rooms) are difficult or impossible to apply in space. One is increasing the mass of the wall or barrier between the two rooms. Because launch weight must be kept to a minimum, simply increasing the weight of the wall is not a desirable alternative. The other obvious option is to dampen the sound by transmitting it outside of the habitat, where it is either dissipated into the atmosphere or ground. This is unworkable in spaceflight environments that do not have an external atmosphere and, in the case of transit vehicles and orbiters, are not anchored to a major mass such as Earth. Foams that have many closed air cells could do a good job of absorbing sound, but vent too many noxious gasses or too easily catch on fire.

Perhaps the most promising method for sound and vibration control are the use of highly viscoelastic materials and the creation of "fuzzy" environments. Viscoelastic materials, such as rubbers and plastics, absorb energy and then dissipate it internally rather than transmit it through the wall or bounce it back in the form of reverberation. For maximum effect, these materials would be put together in multiple layers and loosely joined. Sounds can be further reduced by construction of well-sealed enclosures and sound-reducing partitions placed within the air space of the habitat. Some modern railroad passenger cars, for example, reduce noise by placing transparent glass partitions every few feet within the car. This muffles other people's conversations without reducing the impression of spaciousness.

Some sounds can be very distracting—for example, intermittent sounds or a muffled conversation taking place on the other side of a wall. In certain areas, broadband masking noises can minimize the distractions and annoyances due to such sounds. These include sources of white (broad spectrum) noise produced by fans, or possibly music. Masking noises should either be limited to high privacy areas (sleeping quarters, waste management facilities, and so forth) or delivered by means of headphones so as not to interfere with speech communication.

Personal cassette recorders or CD players are entertaining and help users mentally tune out other people. Personal control over volume and content allows people to enjoy selections that other people may dislike and helps prevent programming conflicts. Used with headphones, these devices do not increase the ambient noise level.

Negative sound systems might work in areas characterized by relatively loud noises that are either continuous or highly repetitive. These systems generate waves that are the complement or mirror image of the waves produced by the annoying noise, thereby neutralizing or canceling the unwanted sound.

Odor Control

Odors accumulate rapidly in confined settings where there is recycled air. Toxic substances have to be controlled for health reasons irrespective of the strength or pleasantness of their odors. Obnoxious but otherwise harmless odors must be controlled to minimize distraction and dissatisfaction. Thus, we must do our best to control undesirable odors, including those associated with personal hygiene and bodily functioning.

Ideally, personal hygiene facilities would allow each spacefarer to wash whenever he or she felt like it and undertake full body cleansing at least twice a week. Spacefarers should have a fresh change of clothes at least twice a week. Garbage should be kept in well-sealed containers, and exposed surfaces cleaned regularly to eliminate contamination that could give rise to odors.

Fortunately, people rapidly adapt to increasing concentrations of odors. People who have occupied close quarters for an extended period of time will not notice the growing accumulation, at least not nearly as much as newly arrived people. Perfumes or other overpowering scents are useless because they increase the concentration of undesirable substances in the air, and in any event, the scents that one person finds pleasant may be deemed unpleasant by somebody else. Thus, people prefer a high rate of air exchange, with "spent" air being decontaminated by a combination of procedures such as electrostatic ionization to remove large particles; dry scrubbing or passing the air through carbon or other chemical beds that trap offending contaminants; and ultraviolet radiation.

CONCLUSION

If we want to provide the strongest possible support for people in space, we must do more than protect them from hazardous conditions and give them the basic necessities to stay alive. We must develop safe, comfortable quarters where people can live and work for extended periods of time. Not only must we satisfy people's needs for life and security, we must also accommodate their needs to relate to one another and to thrive psychologically and spiritually. Arguments that no inconvenience is too great if the reward is the wonders of space, or that all activities are glamorous when they are undertaken in orbit, are based on wishful thinking.

The cost and efficiency of a design may be primary considerations, but they are not the only considerations. Space habitats must be large enough to accommodate a range of human activities, and in determining interior spaces cultural backgrounds must be taken into account. For extended missions, spacefarers require private quarters; semiprivate areas that allow small groups of people to congregate; and public areas such as wardrooms, where the entire crew can be assembled.

Aesthetics are not frivolous and dispensable; they can help ease the stresses associated with isolation and confinement and facilitate high performance. Although under conditions of microgravity "up is where the head is," designs should foster a vertical orientation. Different architectural layouts and color schemes can enhance impressions of spaciousness. Windows are crucial, and photographs of landscapes can "enlarge" the environment psychologically. Real, miniature landscapes with living dwarf trees, along with flowers and other plants, may help.

Good lighting, noise control, and odor control are important for high-quality work. Lighting can encourage certain moods and help break up interior spaces. Background noise must be kept within tolerable levels, especially in areas where astronauts sleep. Ample airflow coupled with air revitalization techniques may reduce the habitat's resemblance to a gym or locker room.

During past migrations from point to point on Earth, notes Wolfgang Preiser, the challenge consisted of cultivating a "rough but basically habitable" natural environment.[44] As we stand poised to move into space the problem is one of humanizing an "artificial, sophisticated, man made environment." Whereas past technological achievements (such as railroads) tended to evolve from existing images (such as horse-drawn carriages), there are relatively few prototypes for outer space, and we can make a fresh start.

CHAPTER 6

SELECTION AND TRAINING

The Boeing Defense and Space Group offers helpful advice for aspiring astronauts. After stating that there are "probably as many paths to that goal as there are astronauts," they identify characteristics that many astronauts share. Most astronauts developed strong interests in science and math very early in school. Most participated in sports and worked during summer vacations, and about a third were involved in scouting. They did well on standardized tests, chose careers in science and engineering shortly after they entered college, and they got great grades. They also learned teamwork—"an absolutely vital skill in the complex undertakings of space research and exploration."[1]

Selection is the process of choosing people who have the greatest potential to succeed at a particular job: in this case, people who have the right technical qualifications to do the work and the right personality and interests to thrive in the unusual environment of space. In the initial phase of selection we may be less concerned with the candidate's immediate suitability than with his or her potential for performing well after thorough preparation. Selection criteria are the standards that are set, and selection procedures are the processes used to assess individual candidates relative to the criteria. NASA selections are conducted by the Astronaut Selection Board, which itself reviews the candidate's qualifications and studies recommendations from other reviewers, including psychiatrists and psychologists.

Right now, only highly selected people fly in space, but this will not always be the case. Space tourism cannot thrive as an industry if too many paying customers are winnowed out. And whenever large num-

bers of workers are needed, as will happen if we construct huge orbiting communities, selection criteria will be relaxed. After the first generation on multigeneration missions, crews will consist of people who
were born onboard. Whereas we hope that heredity will endow subsequent generations with many of the sterling attributes of the carefully
chosen spacefarers who set forth from Earth, some of the people born
in space will be mediocre spacefarers or worse.

Once chosen, candidates must be prepared for their mission. Training helps people develop their potential so that they are at peak read-
iness to perform. More formally, training is "the systematic modifi-
cation of behavior through instruction, practice, measurement and
feedback. Its purpose is to teach the trainee to perform tasks not previously possible, or to a level of skill and proficiency previously unat-
tainable."[2] Spacefarer training takes place in classrooms, in simulators,
in aircraft, and in many field locations.

Over time, as we prepare for each generation of space missions, we
can expect changes in both the kinds of people we need and the kinds
of preparation they will require. The first crew to Mars will consist of
people with the skills to get there, conduct basic research, and get back.
Later on we may need people from many different occupations and
specialties to establish a durable community there.

Selection

Periodically, NASA selects groups of astronaut candidates to become
full-fledged members of the astronaut corps. Typically it reviews approximately 2,000 applications, many of them from people who have
applied before. Of these 2,000, NASA has accepted as many as 35, but
usually somewhere around 20 are selected; that is, 1 percent or less.
These odds are much better than, say, having quintuplets or winning
Power Ball Lotto, but they are still very small.

At present, applicants for the U.S. space program may apply to
become pilot astronauts (this includes mission commander and pilot) or mission specialists (who monitor and operate the shuttle's
systems and manage life onboard). There are also payload specialists who perform mission-specific tasks such as operating specialized
scientific equipment. Pilots and mission specialists are NASA astronauts, while payload specialists come from academia and industry.
For a brief period in the mid-1980s, payload specialists included

representatives of the U.S. Congress, including at least one who played a significant role in NASA's funding. Pilots and mission specialists are career astronauts; payload specialists hire on for individual missions and usually fly only once. Pilot astronauts and mission specialists must be U.S. citizens, but payload specialists have come from many countries, including Saudi Arabia, Japan, and Germany, to name a few. Future international missions may blur the distinction of national origin.

Louis Fogg and Robert Rose traced the fate of 2,288 applicants who participated in the 1988–89 selection.[3] Of these candidates, 179 competed for 7 pilot astronaut positions, and 2,109 competed for 16 mission specialist slots, meaning that there was a higher success rate for the aspiring pilots (about 4 percent) than for the mission specialists (less than 1 percent). Of the 2,288 applicants, 845 lacked the minimum educational credentials and were automatically eliminated, reducing the pool by almost 37 percent, to 1,443. Another 36 applicants who failed to indicate their ethnic or racial group were excluded, although the precise reason for this is unclear. Overall, white men, women, and minorities were hired in proportion to the size of the identifiable applicant pool. However, since white males took all of the pilot astronaut positions, this overall proportionality required hiring a slightly disproportionate number of women and minorities to become mission specialists. In 1992, NASA filled 19 openings for pilots and mission specialists and in 1998 it filled 25 slots.

Although only a small percentage of candidates who complete the paperwork for NASA are selected, it is worth noting that applicants not accepted on the first try may be chosen later on. In some cases the candidates have worked for NASA since their initial rejection and, in their capacity as NASA employees, demonstrated their credentials to everyone's satisfaction. The value of a job with NASA as a stepping-stone to the stars helps account for one of the surprising results of the 1998 selection. After announcing "NASA Selects 25 Astronaut Candidates," the lead-in to an article in *Final Frontier* went on to ask, "Do any live in your home town?"[4] The answer depends a great deal on whether or not you live near Houston! Of the 25 finalists, 8 (32 percent) hailed from Texas: 2 each from Friendswood, Houston, League City, and Seabrook, each of which is within a few miles of Johnson Space Center.

Basic Qualifications

At the time of this writing, candidates for pilot astronauts must be between 162 and 193 centimeters tall and possess a bachelor's degree (or better) in engineering, biological science, physical science, or mathematics. Pilot astronauts must have a minimum of a thousand hours in command of jet aircraft, and experienced test pilots have an advantage. (The best, perhaps only, way to get this experience is as a military test pilot.) Candidates for mission specialist have a little more latitude in terms of height (the acceptable range is 146 to 193 centimeters). They are expected to have degrees in the same fields as pilot astronauts, plus some combination of advanced training and experience in their profession.[5]

Good physical condition has always been a requirement, but today's physical examinations are not as demanding as those detailed in Tom Wolfe's *The Right Stuff*.[6] (In one memorable episode, Mercury astronaut candidates were given barium enemas and then forced to "hold it" while running down a corridor to a room where an examiner took an x ray.) Applicants' blood pressure must not exceed 140/90 in a sitting position, and applicants must have 20/70 vision or better, correctable to 20/20 each eye. Mission specialists have the same blood pressure requirement, but there is a little more leeway in terms of visual acuity (20/200 or better, uncorrected).[7]

Requirements for physical toughness have varied over the years. It was extremely difficult to make the grade during the early years. Remember that when the first astronauts were selected, researchers weren't completely confident that people could survive in space. Spacecraft then were relatively primitive compared to today's, and postlanding recovery procedures (that involved plucking survivors out of the ocean) placed a greater demand on returning astronauts than does gliding home to Kennedy Space Center. Toughness remains important for cosmonauts and their guests who travel on Soyuz, whose return trajectory generates higher G-forces than the shuttle, and whose touchdown may occur far away from the designated landing zone. Russian training equipment is capable of spinning cosmonauts up to 30-G, and cosmonaut candidates have "frozen, shaked and baked" in tiny, completely dark chambers.[8]

"Seed people," writes Marshall Savage, "are necessarily a hardy lot. . . . Paradise has to be carved out of the raw frontier with blood, sweat, and tears."[9] Still, we should avoid making physical fitness criteria needlessly stringent. We should require only that the person withstand the rigors of launch and reentry, do his or her job, and be able to follow emergency procedures. Prosthetic devices should not exclude users from all spacefaring roles unless it can be shown that these prostheses impair the person's ability to perform.

Psychological Criteria

Almost endless psychological qualities could affect a person's ability to work in an exotic setting. This is brought home in the navy psychologist Robert Biersner's discussion of requirements for deep-sea divers.[10] After listing intellectual brilliance, courage, perseverance, social adaptability, humility, and other positive qualities, Biersner concluded that "those who are most suited to work at watery depths should be able to walk on the surface as well."

Not only psychologists and psychiatrists but engineers, newscasters, talk show hosts and guests, science fiction writers, and spacefarers themselves have strong ideas as to what spacefarers should be like. Is there any way to choose which attributes would actually make a difference in space, and which are red herrings that may sound important but are unrelated to on-the-job performance?

The basic validation procedure allows us to determine, in a scientific way, how a particular ability, personality characteristic, interest, or attitude relates to actual performance. This is a three-step process involving (1) assembling a group of individuals and then assessing the strength of a given attribute in each one, (2) having all of these individuals perform in the situation of interest, and then (3) determining if the attribute is statistically related to performance.

Suppose we hypothesize that people who do not take high risks are better choices than those who do—after all, space is dangerous enough without sending people who are always pushing the limits. Next we need a psychological test of some sort to assess candidates' risk-taking proclivities. The test is administered and scores are obtained. However, at this point scores are *not* used for fitness ratings or for selecting spacefarers for future missions. There are several justifications for ignoring the scores, but the overriding reason is that, unless both the person who takes high risks and the person who avoids them get to perform in space, there is no way to determine if risk taking affects how well a person does.

Only rarely has this simple but powerful tool been applied to spacefarers.[11] Whereas we should possess a long roster of traits related to performance in space, we do not. One reason for this is that conducting such research would run counter to the antipsychological bias that is characteristic of NASA. Another is that if the members of a selection board are convinced that they already know what to look for in candidates (even though there is no scientific evidence in support of their hunches), they select people based on their intuition and never find out if they were right. The people who are unfairly excluded never have an opportunity to prove themselves.

Experience in polar and undersea settings suggests a certain broad profile for people who do well under conditions of isolation, confinement, and danger. The precise roster of attributes varies from study to study, but the three basic qualifications identified by Eric Gunderson and others during their psychological studies in Antarctica also apply in space.[12] As summarized by the New Zealand psychologist A. J. W. Taylor, these are "Ability, Stability, and Social Compatibility."[13] Most selection criteria proposed by NASA, the European Space Agency (ESA), the Russian Space Agency (RSA), and others are variations on these three qualifications. They all want spacefarers who are capable of sustained high performance, who are emotionally stable, and who can get along with one another.

Procedures for assessing people's strengths and weaknesses include direct observation, interviews, and psychological tests. Of great concern to some of NASA's international partners is the fairness of the procedures NASA uses to select international crewmembers.[14] Many of the intelligence and other tests proposed for selection purposes were developed in the United States and standardized on Americans. People from other cultures who may be less proficient in English, or less aware of American customs and traditions, could do poorly on these tests, not because they lack the requisite qualities but because the test is biased in favor of native English speakers familiar with North American ways. Even tests that are supposed to be culture-free sometimes handicap people from other countries.

Ability

Aptitude and motivation are crucial. Over the years, selection teams have sought candidates who are self-starting, hard working, and persistent. They seek people who perform well under threat of disaster and keep going despite fatigue, frustration, and hardship. People who have interdisciplinary backgrounds may be of particular interest, and, unlike the military, they may want renaissance people. Consider *Star Trek*'s Captain Picard—musician, actor, anthropologist, and starship commander!

People who are incapable of getting their jobs done or who lack motivation to do them are worse than useless in lethal environments. Crewmembers who work together in dangerous situations demand the best of each other. In the days of sailing ships, commercial sailors signed on ships at very low pay until they gained the experience to earn greater responsibility and higher pay. Different classes of seamen did not take tests or carry licenses or certificates to prove their skills: they simply stated their qualifications when they signed on board. If an avaricious sailor exaggerated his credentials and could not perform at the promised level, he was lucky if he was only ostracized, beat up, and thrown off at the next port. Sailors who made fraudulent claims about their credentials put the whole crew at risk because the ship ended up with too few experienced hands.

Perhaps someday a spacefarer will sign on a tramp starship, exaggerate his or her credentials, and then get pushed out of the air lock by disgruntled colleagues. Today, however, and for a long time to come, spacefaring candidates must take tests to demonstrate their cognitive and intellectual abilities (the specific tests and the weight accorded them have varied). For the most part, today's candidates demonstrate their ability through successfully completing college degrees. Generally they specialize in some sort of hands-on field of science. Despite the fact that for some positions advanced degrees are considered optional, they are very common among astronauts. As far back as the 1960s, even test pilots intent on becoming astronauts earned advanced degrees to increase their chances of selection.

IQ tests measure intelligence in its most general sense. Scores reflect the examinee's ability to comprehend the world, accurately perceive relationships, and solve problems. Although to some observers the early astronauts seemed tongue-tied, they were actually extremely bright. Their overall IQ scores tended to fall into what psychologists call the superior (IQ of 115 to 129) and very superior (IQ of 130 and above) ranges. On occasion, people of average IQ have been selected, but only 10 percent of the astronaut corps has an average IQ, while 48 percent and 42 percent are classified as superior and very superior, respectively.[15]

What will happen as we seek a broader base of skills for space—

when scientific and technical experts are joined by laborers, people from the hospitality industry, space marines, and manual laborers? As we recruit an increasing number of people who would not meet earlier standards of intelligence we will have to make spaceflight a more "user friendly" activity.

Stability

Spacefarers must be emotionally stable in the sense that they are "calm, cool, and collected" individuals with even temperaments. They should neither overreact under stress nor deny the danger in a situation. In the 1950s, notes Brian Harvey, Russians preferred mongrels for animal flights because mongrels tended to be less high strung than purebreds.[16] It seems that some of these dogs were snatched off the street, much as Englishmen once were impressed into His Majesty's navy. Early cosmonauts, Harvey adds, were expected to be "brave, reliable, physically fit, not prone to panic, capable of mental endurance, and familiar with the notion of 'things up there.' "[17] Both U.S. and Russian selection teams try to eliminate people who are easily flustered, who overreact, or whose performance deteriorates markedly under stress.

We also expect spacefarers to be emotionally stable in the sense that they are free of mental disorders. Psychiatric problems are common in our society and affect many people at some point in their lives. Such disturbances include crippling anxiety, confusion, extreme moods, withdrawal, destructive impulses aimed at the self or other people, dependence on alcohol and drugs, and feelings of personal worthlessness. People who have such symptoms find it difficult to get their work done and tend to be socially disruptive. This is dangerous in space, where everyone must be capable of doing his or her job, and where it is difficult enough to maintain one's own equilibrium without having to take care of another person. Successful candidates must be self-controlled, able to tolerate isolation and separation from loved ones, and not have an exaggerated view of themselves.

Paper-and-pencil tests, projective tests that require people to make up stories about vague pictures such as inkblots, and psychiatric interviews are among the tools for assessing emotional stability. Psychiatric interviews are favored by NASA. Until the shuttle era, NASA psychiatrists determined their own line of questioning, but in the 1980s Patricia Santy, at that time a NASA psychiatrist, adopted structured interviews to assess people according to modern psychiatric diagnostic criteria.[18] Since the compendium of psychological problems is a large one, these interviews are fairly lengthy and detailed. Like other job candidates, astronaut hopefuls try to put their best foot forward, so questions have to be worded in such a way as to elicit informative responses. For example, asking people if they are heavy drinkers is likely to elicit a negative response, whereas asking them how many drinks they have had in the past three days makes it a bit more difficult to dodge the issue.

Social Compatibility

Selection teams seek candidates who are compatible in the sense that they can get along with one another under difficult conditions. Social compatibility was irrelevant in solo Mercury missions but gained interest in the two-person Gemini flights. Somewhat incompatible people can perform effectively for a short time, but social compatibility has been an issue on missions lasting a week or two and is almost invariably salient on missions lasting over a month.

Clearly, social compatibility was important in the shuttle-Mir missions, and the astronauts' abilities to relate easily to cosmonauts helped determine the success of their mission.[19] Along with "job competence," "group living" is one of the two dimensions expected to determine spacefarer success on the ISS.[20] Astronauts who show promise are good listeners, considerate of others, and tolerant of both individual and cultural differences.

Research in spaceflight-analogous environments shows that appearance, cleanliness, and general demeanor do make a difference.[21] Under conditions of isolation and confinement, nervous mannerisms and annoying habits are magnified and so is a lack of basic courtesy. The ability to relate to other people is welcome: to understand what they are saying, empathize with them, provide support when they need it, and leave them alone when they so desire. The strength to tolerate frustration and not "take things out" on other people is crucial. Space exploration requires close teamwork, so we are interested in candidates who want to excel but who can cooperate with one another. People who enter space may earn acclaim while they are there, but people who sign up because they are starstruck and hope to become famous can rub their colleagues the wrong way. Similarly unwelcome are crewmembers who are power hungry or who like to throw their weight around, perhaps by directing a steady stream of instructions and criticisms at others.

Other bases of compatibility are found in the ways that the different crewmembers' characteristics or traits fit together. People who share basic values and attitudes get along well with one another. This does not mean that the different crewmembers have to be like so many peas in a pod, only that there is fundamental agreement on basic issues such as the conduct of the mission. In other cases, people's differences mesh together in such a way as to make it easy for them to get along. The Russians, for example, have paired dominant and submissive cosmonauts, on the grounds that two dominant cosmonauts would fight with one another and two submissive cosmonauts would have trouble reacting in prompt and decisive ways. The dominant personality is named commander.

TRAINING

Astronauts must learn not only the basic tasks of pilots and mission specialists but also how to conduct themselves in ways considered appropriate for astronauts: how to express the right kinds of attitudes and values, maintain the appropriate degree of emotional control, and represent their space agency in the ways that their space agency likes to be represented.

In a sense, preparation begins in broad form before a candidate joins a space program. We know from studies of people in many lines of work that learning occurs prior to entering an occupation. Before showing up for an interview at Johnson Space Center astronaut hopefuls have already developed a picture, if not entirely accurate or complete, of what it is like to be an astronaut. From reading about astronauts, hearing astronauts speak, and watching documentaries and even science fiction films they have some ideas about how they should act, just as premed students' experiences with real-life and television doctors allow them to imitate physicians before they enter medical school. Some of today's children who participate in mission simulations conducted by the Challenger Centers and various space camps are learning, in rudimentary form, a few of the skills required to become tomorrow's spacefarers.

Informal and Formal Training

Preparation has both informal and formal components. Informal learning comes from interacting with experienced astronauts, having off-the-cuff conversations with instructors, and discussing matters with one another. Candidates learn such subtleties as the need to hide personal ambitions, the kinds of assignments that are likely to lead to an actual flight, and the necessity to "suck up to go up" and to say, "Houston, we have a problem," rather than use a four-letter expletive after something goes dreadfully wrong.

Formal training programs are offered at both beginning and advanced levels.[22] In the U.S. space program, astronauts spend a year as "astronaut candidates" in basic training at Johnson Space Center before becoming astronauts eligible for assignment to a mission. They are required to take water survival training and become SCUBA qualified in preparation for space walks. They gain experience in high-pressure and low-pressure chambers and learn how to deal with altitude shifts during emergencies.

Preparation includes course work and computer-based lessons in astronomy, mathematics, physics, geology, meteorology, and oceanography. Candidates learn also about spaceflight and spacecraft, including orbital dynamics and navigation, propulsion and environmental control systems, and space suits.

Next comes intensive work on single systems trainers (SSTs). Each SST emulates one of the shuttle's systems. Guided by an instructor, and with the use of checklists, astronauts learn how to run the system, detect errors, and correct them. This is followed by training in shuttle mission simulators (SMSs), which involve all aspects and phases of shuttle operations.

Integrated training includes the flight controllers, flight surgeons, and other support personnel who work with the crew in space. As a result of training with the space crews, support personnel learn the crewmembers' strengths and weaknesses, can anticipate the crewmembers' needs, can communicate effectively, and can tell when a crewmember is under stress or something is amiss. There is enough to attend to in space without having to work with support personnel who are unknown quantities or who have yet to be brought up to speed. If all goes well, there develops a strong sense of mutual trust.

The importance of trust between workers in the field and support personnel is evident in the recent experiences of a state police agency. The barracks are located throughout the state, and for years each barracks had its own radio communications post. The radio operators and the officers knew each other well and were, in effect, part of the same family. Personal friendships strengthened the operators' determination to provide the officers with the best possible support, and the operators developed a "sixth sense" about the officers' welfare in the field. Management switched to a system that removed radio operators from the barracks and concentrated large numbers of them in huge, centralized radio communications posts, which meant that by and large the operators and officers were working with people they did not know. For some officers, working with highly trained and well-meaning but anonymous operators added new layers of uncertainty to a job already difficult enough.

Astronauts' advanced training involves preparation for a specific mission. This may consume fifty to sixty hours a week prior to liftoff. Astronauts are likely to undergo ten months of training before each mission and more for a satellite rescue or international flight. Training involves using software and procedures developed for the mission, During the last few weeks, it includes working with mission control.

In the Russian program, the longer the anticipated mission, the longer the crewmembers train as a team. Their normal training for an extended-duration international mission may exceed two years. Bryan Burrough believes that some of the problems that astronauts encountered on Mir were due to the short amount of time that the flight and ground crews had to get prepared.[23]

Training to live together under prolonged conditions of isolation and confinement has received greater emphasis in the Russian program, which has required cosmonauts to remain aloft for many months at a time. For example, in the early Russian program, some cosmonauts were sent on a one-month automobile tour of the Soviet Union, cramped together in a small car, to see how they got along with one another and to give them a taste of their future life together. Similarly, cosmonauts have been confined in isolation chambers for up to a year. Subjecting people to prolonged periods of isolation and confinement in polar regions, hyperbaric chambers, and deep caves has been characteristic of the European Space Agency, but not of NASA.

In the U.S. program, training accelerates as launch time draws near. The program might, however, temper its tendency to increase training just prior to departure with recognition that training becomes a liability if it becomes so strenuous at the last minute that the crew is already exhausted at time of liftoff.

Applying Principles of Learning

Centuries of experience in the military, in industry, and in the classroom reveal what we need to do to develop successful learning programs. First, we must walk before we can run. It makes sense to have students master a graduated series of tasks, starting with those that are relatively easy and then moving to those that are more demanding. The key is to set difficulty levels high enough to be challenging, but not so high as to make the task impossible. Exercises should be designed to stretch but not exceed the learner's capabilities.

We learn through observation. More experienced spacefarers serve as mentors and role models for the less experienced spacefarers. However, in a group training situation, experienced spacefarers should know when to stand back so that the newcomers can demonstrate initiative and gain experience.[24] Spacefarers also learn through providing support for one another's missions. During this they see how other crews manage their mission and handle the various glitches or problems that arise.

We do best when we receive prompt, accurate, and detailed information on how well we have done. We need to know what we did right, what (if anything) we did wrong, and what we must do to improve our performance the next time. Prompt, continuous feedback is crucial to a spacefarer's training.

Especially when it comes to responses that must be performed flawlessly under a wide variety of conditions, overlearning helps. That is, even after we have learned how to do something right, continued rehearsal is desirable. This is because performance levels that seem perfectly acceptable during practice may deteriorate rapidly when the chips are down and the person is working under stress. Riding a bicycle is a good example of the benefits of overlearning. People who have not bicycled for decades usually discover that it takes only a very few minutes of practice to regain proficiency. As children, they had hundreds of hours of practice, so although their bike-riding skills may get a little rusty they are never really lost. For this reason, spacefarers practice critical tasks dozens, perhaps hundreds, of times.

We learn best when practice sessions are distributed over time. If we want to learn the lines for a play, for example, we are better off reciting them at different times and under different circumstances than trying to acquire them all at once. Even if it were possible to memorize everything, recollection is not as good as when learning is spread out over time.

In spacefaring, considerable time may pass between the time a skill is learned and when it must be used. If a mission to Mars takes several months, then landing skills that had been carefully honed on Earth may become rusty or imperfect. One of the reasons advanced for a Progress resupply vehicle's crashing into Mir was that the commander, who was using two joysticks to ease the ferry into the dock, had not practiced the maneuver for several months.[25] Thus, it is useful to hold practice sessions in space to keep skills fresh. On the shuttle, astronauts use computers to simulate cockpit activities before they return to Earth. Their laptop computers will make it possible to practice skills on the ISS and on a mission to Mars.

Finally, we do best when we start with individual elements or components of a job and then combine them into an organized sequence. When learning how to play a musical instrument we may first discover how to play the melody and then how to add chords. Later, we play both the melody and the chords together. Similarly, astronauts first learn individual skills, then combine them into complex patterns and sequences, and then integrate them with the activities of other crewmembers. H. S. F. Cooper Jr. quotes astronaut John Young as stating that crystallization, or coming together as a team, "happens after about a thousand hours of training—usually a month or two before launch." When that occurs, "a crew of three can do the work of a crew of four or five" and the team seems able to "solve problems automatically, without much talk."[26]

Simulators

Realistic training situations increase the learner's enthusiasm and ease the transition from make-believe to real. Mock-ups bear a physical resemblance to a space vehicle or habitat, but the resemblance is not highly defined and, like a clay model of a preproduction automobile, there are no moving parts. Mock-ups are useful for checking architectural designs—such as verifying that the interior space allocated to a compartment is sufficient for the activities that must take place in that compartment—and, when mock-ups are put on display, for public relations purposes. On one mock-up of the ISS the various hatches, compartments, windows and other fixtures are represented by simple painted outlines.

Simulators, on the other hand, are realistic. Controls and gauges closely mimic the real thing: buttons can be pushed, switches thrown, and dials twisted, and the gauges and video displays respond. Simulators give convincing hands-on experience. Simulators range from relatively simple devices such as a control panel and display for a single system to full-scale simulators that involve all crewmembers working on all systems. Simulators do not necessarily have to be highly realistic for us to learn. Nonetheless, modern aircraft simulators are so realistic that time spent operating a simulator substitutes for time spent flying a real aircraft, and a pilot's performance is evaluated by having a check pilot observe him or her take the simulator on a make-believe flight.

Johnson Space Center has two shuttle simulators. One of these, the stationary, full fuselage trainer, has a cockpit, crew quarters, and a cargo hold that closely match those on the actual shuttle. This model of the shuttle is highly detailed inside and out and includes storage compartments, air-scrubbing cartridges, and everything else.

The second simulator, a crew cockpit simulator, bears less external resemblance to a shuttle in that it simulates the cockpit and crew quarters alone, omitting the fuselage. However, the cockpit is precisely detailed and the model itself moves and responds as the crew operates its controls. Inside this craft, learners manipulate the controls, feel the shuttle respond, and peer out the windows at video displays that mimic sights seen while taking off, orbiting, and landing. Astronauts report that the effects are very realistic and lack only vibration and weightlessness when compared to the real thing.

During simulations, astronauts deal not only with routine conditions (to the extent that anything is "routine" in space) but also with various unexpected contingencies, including emergencies. Trainers arrange for various system failures, sometimes one right after the other in very short order. One of the differences between the simulators and the actual shuttles, notes Cooper, is that in the simulators the abort switches and other emergency controls are well worn. "It was proverbial," he stated, "that a spaceflight was a snap compared to a session in the simulators."[27]

Modified corporate jets called shuttle training aircraft give astronauts experience with shuttle landings. Templates block out part of the windshield so that the view closely approximates that from the shuttle's windshield. During landings the jets are thrust into reverse to provide the appropriate feel. This gives the pilot the opportunity to practice landing an aircraft with the shuttle's handling characteristics, and both the pilot and passengers learn how to acclimate to the very rapid rate of descent.

There are even ways to acquaint spacefarers with microgravity before they enter space. One method, mentioned earlier, is parabolic flight. While a jet airplane known as the "Vomit Comet" flies a parabola (a sharp, inverted U) passengers experience up to thirty seconds of microgravity. Both U.S. and Russian programs have large jet transports with padded interiors to ensure that nobody gets hurt when they are reclaimed by gravity after a brief period of free fall.

Astronauts are further conditioned to microgravity in neutral buoyancy chambers at the Sonny Carter Training Facility. This consists of a 23-million-liter pool that is 62 meters long, 31 meters wide, and 12 meters deep. Facilities like this are used for engineering evaluations as well as for training flight crews. For example, in an underwater mock-up of the ISS, engineers can determine the proper location or placement of foot restraints, hand rails, and other devices. They can evaluate the design of various pieces of equipment and tools in a neutral buoyancy environment.

Astronauts don diving suits (which are analogous to space suits) with weights attached and then enter the tank. The trick is to achieve a delicate balance of forces so that the trainee neither sinks to the bottom nor rises to the top but is essentially able to move around in three-dimensional space much as he or she could move around in microgravity. Neutral buoyancy training is useful for practicing space walks and gaining a sense of what it would be like to construct something in orbit. At present, astronauts undergo about ten hours of neutral buoyancy training for each hour of extravehicular activity anticipated. This number will be reduced in the future as work in the space environment proliferates and extravehicular activity becomes more common.

John Schuessler notes that neutral buoyancy training is much more difficult than operating in a space suit in microgravity.[28] The weights make the simulated space suit buoyant, but the astronaut inside the suit is not neutrally buoyant. The suit is inflated in such a way that it tends to surround the astronaut like a balloon. The astronaut inside tends to sink to the bottom of the inside of the suit in whatever position it is oriented as gravity pulls the astronaut's body downward. That means that the astronaut is lying on one surface inside the suit, which is a very difficult position in which to work. Neutral buoyancy training is good because, once completed, it is easier to do the same task in space.

EDUCATION IN SPACE

In the future, spacefarers may establish entire school systems to educate their children and keep their own knowledge fresh. In near-Earth orbit, and even on the Moon or Mars, these could in many ways resemble Earth's educational systems, although somewhat reduced in scale. At first, such systems will have to be abbreviated, but they will gain strength as the community matures.

This poses some interesting challenges for educators. It may not be possible at first to sustain a cadre of professional teachers. Parents and other adult spacefarers will have to serve as teachers, role models, and mentors. The learning situation or classroom may resemble that of a home learning situation of today. Because the spacefarers are likely to be technically oriented, it will be challenging to ensure a broad representation of topics or subject matters (including the arts, humanities, and social sciences). When the NASA-contracted educator Don Scott asked the astronaut John Young, "What should I teach students who are possibly going to Mars?" Young replied, "Math, science, engineering, and medicine. And poetry—we will need poets on Mars."[29]

A liberal arts education may seem superfluous, but it feeds the soul and fosters the values and skills useful for community leadership. As people enter positions of prominence, their specific technical skills become less relevant and their basic abilities to think, make value judgments, and reach decisions gain in relevance.

We can expect that there will be heavy reliance on computer technology. This could include computer data banks containing the same books and videos that will be available in educational institutions on Earth. By that time, people may be thoroughly acclimated to learning from interactive compact discs and taking virtual classes on the Internet. These may be common on Earth as in space, so there may be no specific handicaps associated with obtaining a degree from Old VU (Virtual University). But there remains the issue of injecting new knowledge from Earth into the curriculum. Laser communications may allow the infusion of massive amounts of information from Earth, but real-time communications delays, eventually measured in light-years, will make it difficult to school people in current affairs, at least the way we think of them on Earth today. The spacefarers' children may be like children in isolated parts of the United States, such as Appalachia in the 1930s and 1940s. Essentially cut off from the rest of the world, they spoke their own quaint forms of English and were oblivious to events in other towns.

Training new generations of physicians and other professionals may be particularly challenging.[30] Interstellar travelers, for example, will not have the luxury of selecting the top undergraduates from Earth's best colleges. They will have to select from a relatively small number of applicants who were born in space and received their basic education there. Furthermore, the "medical school faculty" is likely to consist at first of one or two medical personnel who accompanied the initial settlers. Indeed, these people may be generalists—biologists with some training in medicine, or practitioners who combine medicine, veterinary medicine, dentistry, and optometry. How, under such circumstances, can we hope to develop an effective program to train other generalists, never mind the specialists who will be in demand as the colony grows? Developing professional schools in space is not a pressing concern, but it is one that we may have to address at some future time.

CONCLUSION

There is a similarity among the astronauts who have been chosen over successive generations of missions. Perhaps the description of the criteria for early astronauts framed by Kubis and McLaughlin does not differ all that much from the qualities sought today: the program sought candidates "with a high degree of intelligence[,] . . . an ability' to work closely with others[,] . . . flexibility and adaptability to meet any emergency without psychological disintegration . . . [, and] an outstanding capacity to tolerate stress.[31] In this description, written over thirty years ago, we find ample allusion to ability, stability, and social compatibility.

One reason for this similarity across generations is that once we have developed conceptions or stereotypes of what it takes to become an astronaut—and in NASA these conceptions developed early on—it is difficult to shake them. Furthermore, there may be a certain momentum brought on by tradition. We know also of rating errors in personnel evaluation called "similar-to-me effects." These involve favoring people who are similar to ourselves, or at least assigning them evaluations higher than those we assign to people who are dissimilar to us. Perhaps spacefarers involved in the selection process would be drawn to candidates who share their talents and interests. Thus, by choosing people who fit their own generation's image, each generation of spacefarers tends to perpetuate itself. The resemblance is further sharpened as the older generation mentors and coaches the younger generation.

We can expect problems, however, if selectors keep using the same cookie cutters, stamping out the same spacefarers, as mission requirements change. While we will always want people who are technically expert, we will need people with expertise in more and more areas. The strong quantitative and spatial skills that have typified spacefarers over the generations may become less important as artificial intelligence and automation improve, and as new jobs in space open up. Similarly, we will always want people who are even-tempered and emotionally stable, but the requirements may become less stringent as habitats become more comfortable and we find new ways to ease adaptation to space. Social compatibility will always be important, but the skills required to get along with other test pilots are not necessarily the skills required to get along with everyone else.

We must always be sensitive to changing conditions. Are we insisting on people who have skills that are no longer—or never were—relevant? Are we training people in areas that no longer—or never were—pertinent? If we can answer yes to either of these questions, we are falling down on the job.

Training programs have been remarkably successful, at least if we focus on the aspect of spacefarers reaching their objectives and coming back alive. At the same time, training too must be revised to reflect changing mission requirements—not only the newly introduced technology but also the shifting social and cultural climates. We have strong and perhaps mostly valid ideas about what it takes to select and train people for space, but we must continually reevaluate them.

CHAPTER 7

STRESS AND COPING

The Apollo Moon landings are arguably the most spectacular achievement in the history of humankind. On July 16, 1969, when Neil Armstrong, Buzz Aldrin, and Michael Collins set forth for the first lunar landing, nobody could be sure that Armstrong and Aldrin would be able to successfully ascend from the Moon's surface and rejoin Collins, who was orbiting above. The president at that time, Richard M. Nixon, had prepared a brief announcement in case Armstrong and Aldrin were lost. He wrote, "Fate has ordained that the men who went to the moon to explore in peace will stay on the moon to rest in peace. These men, Neil Armstrong and Buzz Aldrin, know there's no hope for their recovery. But they also know there is hope for mankind in their sacrifice."[1] On the thirtieth anniversary of their departure, Armstrong recalled that he had a gut feeling that there was a 90 percent chance that the trio would return to Earth, but he was less confident that he and Aldrin could actually complete the Moon landing phase.[2]

The possibility of not returning alive is only one of many stressors that act upon spacefarers. Stressors are the environmental or other forces that place demands upon people, sometimes overtaxing their physical, intellectual, and emotional resources. Stress refers to people's responses to these demands. Stress is reflected in physiological states, emotions, and activities. Often, but by no means invariably, stress affects people in adverse ways.

SOURCES OF STRESS

Spaceflight subjects people to a wide range of physical, psychological, and social stressors. A recurrent concern is that these stressors could add up in such a way as to undermine a spacefarer's performance. The results could be negligible—for instance, it might take a little longer to prepare a meal than normal, or it might require two tries rather than one to load film into a camera. But the results could be profound if a highly stressed spacefarer bungled a docking maneuver or left a landing module tipped on its side with no way to right it. Despite our fascination with spacefarers and other intrepid people who seem able to weather all sorts of difficulties, we need to limit their stress to assure their survival, make their lives a bit easier, and protect our investment.

Physical Environmental Stressors

Over the years, spacefarers have been tested on rocket sleds and centrifuges that built up tremendous G-forces, shaken on vibrating chairs, cooped up in tiny structures, baked and frozen in special chambers, and taught to survive in the desert and on the sea. Launch and reentry are highly dangerous, and even though some spacefarers may report little stress during launch, physiological indicators such as heart rate sometimes suggest otherwise.[3] Spaceships are, after all, rockets with hundreds of thousands of parts that rely on the controlled combustion of explosive chemicals for moving around, and, as people discover on occasion, there is a thin line between combustion and explosion.

Inside their habitat, spacefarers are subjected to cramped and crowded living conditions, equipment that is difficult to operate or requires high maintenance, inadequate illumination, the constant din of life support machinery, unpleasant odors, and exhausting work schedules perhaps interspersed with periods of boredom.

Spacefarers have constant knowledge that their habitat or space suit could fail (subjecting them to vacuum, temperature extremes, and other hazards) or that they could run out of fuel or other crucial supplies. The structural integrity of the habitat and the reliability of the life support systems are particularly worrisome. During the shuttle-Mir missions, maintaining sufficient atmospheric pressure and adequate oxygen were major concerns; Elektron, the major air revitalization system, was forever breaking down.[4]

Spacefarers are also aware of invisible threats, although these are not always in the forefront of their minds. These include the biomedical effects of spaceflight, such as cardiac and muscular deconditioning, osteoporosis, and an increased risk of cancer and other problems due to radiation exposure. Because of their invisibility, their presence, progress, and effects are difficult to gauge. For some people their ambiguity may arouse more apprehension than do dangers that are easy to monitor. Under conditions of uncertainty we find it more difficult to take effective self-protective steps.

Spacefarers know that because they are separated from Earth they could have to wait weeks to be rescued. Even if an emergency "lifeboat" such as a Soyuz transit is nearby, there may be strong pressures not to use it. Spacefarers on a Mars mission will be effectively cut off from resupply, and there will be few or no prospects of rescue if something goes wrong. Events such as fires that are dangerous on Earth are even more dangerous in space because there is less hope for rescue.

Also, space crews in flight have little or no control over many of the events that directly affect their welfare. For example, during the Cold War, some people in Antarctica feared that they might be left there forever in the event of a nuclear war. Russian spacefarers who were in orbit during the dissolution of the Soviet Union stayed aloft until the political situation in Russia calmed down and they were able to undertake a normal landing.[5]

Interpersonal Stressors

Two of the greatest potential stressors associated with life in space are isolation and confinement. First, people are cut off from society, perhaps by vast distances—in the case of a lunar voyage hundreds of thousands of kilometers. Second, with the exception of solitary missions, spacefarers are forced into close proximity with one another. This paradoxical combination of too much and too little distance from other people is a defining characteristic for spaceflight and spaceflight-analogous environments.

The most obvious challenge of isolation is separation from the people who really matter: spouse, parents, children, and best friends. These are the people to whom we look for guidance and emotional support, and whose reactions to us have the most profound effects on our sense of self and well-being.

Even during training and preparation for a mission, spacefarers are seldom home. Early astronauts spent a lot of time on the road, moving between different bases, facilities, and training sites and putting in countless hours with contractors who were developing equipment for their mission. Today's spacefarers may be sent abroad for months, even years, to become proficient in a foreign language or to train with an international crew, and it is not always possible for their families to join them.

Many people have workaholic tendencies, spend too little time enjoying the children as they grow up, and leave things to the spouse to manage. But only a few jobs, such as submariner, polar explorer, and astronaut, subject people to such high degrees of separation. Astronauts with military experience report that separations due to life as a professional astronaut are more daunting and troublesome than separations associated with military field assignments.[6]

People who are cut off from their families experience a certain helplessness or guilt from not knowing what is going on and from not being able to take action. Thus, it may be even more devastating than usual to learn of the terminal illness of a parent or receive a "Dear John" communication that terminates a relationship. The situation has eased somewhat with improved communication—today, spacefarers can communicate fairly regularly with people on Earth via protected radio channels and e-mail—but it will be difficult or impossible to keep in touch when we embark on interplanetary and interstellar missions.

Confinement offers additional challenges. Crowded living conditions are themselves stressful. When we interact with the same people without relief day after day after day, their ideas and stories get stale. In some missions, rivalries and jealousies crop up, and in outer space it may be difficult to repair relationships that have fallen apart (see chapter 8).

Responsibility for other people may contribute to the "commander's syndrome," which involves crew leaders obsessing over their duties. Additional stress comes from shame, that is, the belief that one will seem unworthy or a failure in the eyes of others. To save face, spacefarers may be prompted to take unnecessary risks or "go down with the ship" to avoid censure. This is magnified by knowledge that one is living in a fishbowl and that every mistake will be reported and perhaps sensationalized by the media.

A study of Australians who wintered-over in Antarctica underscores the importance of social stressors.[7] After they had been in Antarctica for a while, participants ranked various stressors in terms of impact on themselves and on other expeditioners. In three separate samples, they rated social stressors as the most severe. The first-ranked stressor, for all three samples, was separation from family and friends. Lack of privacy, living and working with the same small group, and pressures to conform to the wishes of the majority were close behind. Stress imposed by the physical environment was ranked near the bottom by all three groups. Before they went to Antarctica the participants did not expect separation to be of much consequence. It was only as their mission progressed that they recognized the stressfulness of the social environment.

Organizational Stressors

When we think of the risks of police work, typically we imagine high-speed automobile chases and accidents, shoot-outs, and other dramatic events. Such events do happen, and they have a profound effect on peace officers' lives. Yet studies of police work suggest that confrontation with criminals is only one source of stress. The other is the way that some police agencies are organized and managed. Commonly reported stressors include a lack of backup from higher administration, unsympathetic managers, little opportunity for promotion, exhausting and ill-conceived work schedules including double shifts, and hour after hour spent waiting to testify in court.[8] For many officers, it is not the infrequent high-speed chases and shoot-outs that are the most stressful, it is the accumulation of stress from organizational forces.

We can only speculate on the organizational stressors that could befall contemporary spacefarers. Certainly ambiguity surrounding assignments is stressful: not knowing when or even if one will be assigned a flight, and once the flight has been assigned, not knowing if there will be some sort of last-minute substitution. If the spacefarer believes that he or she has no control over the situation—that assignments are arbitrary or based on favoritism or political considerations—then the situation is likely to be worse. Another stressor is a sense that one must be on one's best behavior all the time, never really questioning or contradicting management, and never expressing negative emotions in public. This would be powerful if one were afraid to speak one's mind on a critical issue, such as safety, or believed that rightful concerns would be ignored. And it is stressful to be held responsible if something goes wrong on the mission, even though the mishap was a direct result of something else, such as faulty engineering or poor judgment on the part of a manager.

CONSEQUENCES OF STRESS

Stress is reflected in people's moods, emotions, and health—among other things. Yet as we look at stress in spaceflight and spaceflight-analogous environments the picture that emerges is not entirely consistent. Conditions that seem like they should be tremendously stressful do not always have adverse consequences. Desmond Lugg found a low incidence of psychopathology in Antarctica, with some national expeditions reporting no problems at all.[9] Benjamin B. Weybrew and Ernest M. Noddin have shown that submariners score exceptionally well on standardized tests of psychiatric functioning.[10] Submariners tend to show higher levels of mental health than do surface sailors, who are in turn better adjusted than members of society at large.

Indeed, conditions that would strike some people as stressful sometimes have positive, or salutogenic, effects. Spacefarers interviewed by Frank White reported that, despite occasional disorientation or other problems, they enjoyed their visits to space and that their journeys had profound and predominantly positive effects on their lives.[11] Donna Oliver reports that several of her Antarctic winter-over colleagues considered the experience a pleasant "return to a simple and more enjoyable life-style."[12] Although they did not deny occasional physical and psychological discomfort, most rated the experience as "one of the best of their lives" and were interested in wintering-over again. Responses to psychological tests showed that as the austral winter progressed, participants became less dependent on others, less paranoid, and more existential, and they gained an increased capacity for close interpersonal relations. A recent study by Lawrence Palinkas and his associates found evidence of reduced tension-anxiety, fatigue, confusion, and depression during a three-week expedition to Isachsen, an abandoned and decommissioned Canadian weather station in the Arctic.[13]

Vyacheslav I. Myasnikov and Iltuzar S. Zamaletdinov report that psychological tests administered two to three years after a space mission reveal improved psychological functioning after spaceflight.[14] Following flight, cosmonauts were more realistic, had greater emotional stability and self confidence, and were more accurate in their self-evaluation. They were less anxious, hypochondriacal, and depressive, less likely to engage in expedient behavior, and less likely to direct aggression toward other people. These findings too suggest that isolation and confinement have salutogenic effects.

Remember: in interpreting people's reactions we must take into account not only the environment but also the nature of the occupants and the conditions of occupancy. These findings represent only some combinations of settings and people. Individual differences in the ways that people perceive and interpret situations are one reason for loose coupling between seemingly stressful conditions and adverse reactions. As Peter Suedfeld points out, the same conditions that one person finds frightening another person may find challenging or even fun.[15] Preparing to set forth on a perilous around-the-world balloon flight, pilot Dick Rutan was asked why anyone would want to undertake such a dangerous mission. "Why would somebody not do this?" he replied. Rutan, who had already flown an airplane around the world without stopping or in-air refueling, explained that "nothing worthwhile was ever accomplished without some kind of risk, some kind of danger."[16]

Another reason for the loose link between seemingly stressful conditions and adverse reactions is that people have many ways to handle stress, limit its effects, even turn it into something positive. Under stress, they may develop skills that remain useful later in life. Palinkas did a follow-up study of U.S. Navy participants in Operation Deepfreeze (1958–59) and found that wintering-over had *beneficial* effects on health.[17] His study involved sailors who volunteered and qualified for wintering-over in Antarctica. As a result of changes in orders, some of these men actually wintered-over whereas others did not. Follow-up over a period of years revealed that men in the winter-over group experienced *fewer* illnesses and hospitalizations than did those who had been reassigned to easier duties. One possibility is that they developed new coping skills that not only helped them weather their period in Antarctica but also carried over to the rest of their lives, thus accounting for their subsequent good health. Studies that have tracked astronauts' health over the years show that they, too, tend to be healthier than people who matched them in terms of age and a few other selected characteristics but who had never experienced the rigors of space.[18]

Cognitive Effects

Reports from polar regions, under the seas, and outer space suggest that intellectual tasks become more difficult in these places than in normal environments, and that inefficiencies and errors creep in. Commonly reported problems include decreased alertness, loss of concentration, memory failure, and perceptual distortions. One early study found that, after several months in Antarctica, thirty out of eighty-five participants experienced absentmindedness and wandering attention.[19] In a few cases the reaction constituted a fugue state. A person might "come to" far from his quarters and have no idea how he got there.

Summarizing more than two decades of research, Arreed Barabasz wrote that isolation increases suggestibility and hypnotizability.[20] Some people become impervious to distracting events and increasingly immersed in their own thoughts. Marianne Barabasz found that in Antarctica people became so engrossed in activities such as reading that other people might have to call their names two or three times or even hit them to get their attention.[21] Her research subjects reported that they could "block things out." One stated that he "could make my own stories, live them in my mind as if they were real life," and another that "when reading books on telekinesis, I felt almost like I had the power to get out of here."

Although absorption and involvement might make it difficult for a person to respond promptly and effectively in an emergency, there are some positive aspects. In the course of looking inward, people tap the wellsprings of their imagination.[22] Anthony Storr proposes that solitude and isolation throw people on their own resources.[23] In the course of learning how to make do on their own, they develop imagination, talent, and creativity. For many writers and artists, isolation eliminates distractions and allows time to create. Simultaneously, unusual conditions yield good material for incorporation into artistic creations.

Health

Stress is involved in anywhere from 50 percent to 70 percent of physical illness, including degenerative diseases such as heart disease and stroke, ulcers, headaches, diabetes, cancer, and infectious diseases, including both viral and bacterial infections.[24]

People in spaceflight and analogous environments tend to have high rates of health complaints. Headaches, nose and chest colds, and digestive problems such as upset stomach and constipation are common within underseas, polar, and outer space environments. The former chief astronaut Deke Slayton believes that stress was responsible for coldlike symptoms that appeared in three successive Apollo missions; one reason for keeping astronauts in quarantine prior to a mission is to decrease the number of people who are placing demands upon them and thus to alleviate stress.[25] The Russians go even further, sometimes sending cosmonauts to a spa for a few days so that they are relaxed and fresh for liftoff.[26]

Astronaut William Pogue reports that a malady known informally as the "space crud" was common among Skylab astronauts.[27] This he describes as a "general malaise or 'down' feeling that is similar to onset of flu or a cold and is quickly relieved by eating." Psychiatrists suspect that some bodily symptoms reflect "somaticizing," or translating psychological problems, which are personally and socially unacceptable, into physical problems, which are not only acceptable but may elicit concerned reactions from other people.

Sleep disturbances, sometimes serious sleep disturbances, are common in unusual environments. Launch day is very strenuous, and spacefarers sometimes become very tired a few hours into the mission. On some missions, cosmonauts have slept so soundly that ground personnel wondered if they were still alive. After twelve days in space, one Russian crew had to be woken up with a siren.[28] Stress, depression, and exhaustion are associated with drowsiness, napping, and long periods of sleep.

Instances of anxiety and depression are common. Research during Operation Deepfreeze found that early in the mission approximately 30 percent of the men reported feeling "blue," a percentage that increased to around 40 percent during midwinter and then leveled off.[29] Palinkas's debriefing of people who had wintered-over found that, despite the improvements in living conditions, about half of the respondents reported sadness and depression.[30] Another study led by Palinkas showed increased depression as the winter progressed.[31] Otherwise normal people lost their appetites and suffered weight loss; had trouble sleeping or slept too much; felt "slowed down" or had trouble moving; had trouble concentrating, thinking, and making decisions; felt resentful, irritable, or angry; and were more in need of reassurance or help.

Brian Harvey states that one Russian mission commander collapsed under stress, and, on the basis of interviews, Bryan Burrough concludes that psychological problems (mood, performance, and interpersonal issues) caused Russian ground controllers to terminate three missions over a twenty-year period.[32] Certainly Burrough's account of the shuttle-Mir missions have sensitized us to the potential stressfulness of life aboard a space station and the occasionally deleterious effects of stress on crew performance and welfare.

Psychiatrists Nick Kanas and Patricia Santy point out that if we send enough people to space for long enough periods of time we will see some serious mental health problems.[33] Certain disorders, such as schizophrenia (an illness characterized by disorganized, perhaps bizarre, thought patterns), typically have their onset in childhood or young adulthood, long before people are at the age where they might be eligible to fly in space. Until we enter an era of space settlement or multigeneration missions, we are more likely to encounter problems that tend to appear in adulthood. These include affective or mood disorders, such as depression, or bipolar disorders, where the person swings between periods of depression and elation. We can also expect some anxiety disorders (vague but sometimes intense feelings of apprehension and dread) as well as strong, focused fears, known as phobias, and occasional panic attacks.

Some people, adds Santy, may develop a fear of spaceflight, just as many other people, including some aviators, develop a fear of flying. Other potential problems include antisocial or deviant behavior, such as substance abuse, flagrant irresponsibility and dishonesty, and violent behavior known as "acting out." Additionally, she states, we must be alert to the behavioral effects of toxicity; that is, the accumulation of dangerous substances in the cabin's atmosphere (due to such problems as outgassing or the failure of air revitalization equipment), which can cause forms of poisoning whose first symptoms include derangement and bizarre behaviors. Usually, it is the person deprived of oxygen who is the last to realize that something is amiss.

Psychological Reactions over Time

J. H. Rohrer studied fluctuations in submariners' mood and morale over time and found three stages.[34] The first stage, which began as the mission got under way, was accompanied by heightened anxiety. The second stage consisted of settling down to routine. This was marked by depression. The third stage, one of anticipation of the end of the mission, was marked by emotional outbursts, aggressiveness, and rowdy behavior. This pattern is not invariably found, but it has been found in Antarctica, in submarines, in space simulators, and in fallout shelters.[35]

Robert Bechtel and Amy Berning point out that this pattern is very common in isolated and confined groups.[36] In Bechtel's studies of military forts and air bases in the far north, for example, base commanders, chaplains, schoolteachers, police officers, and other professionals who work with people reported that the hardest part of the winter seemed to be in February, *after* the peak of winter cold had passed. Their analysis suggests that difficulties peak during the third quarter of a mission, hence they propose calling this the "third quarter phenomenon." What is particularly remarkable is that these stages depended on the relative, rather than absolute, passage of time. In other words, it doesn't matter if the mission is four days, four weeks, or four months, the stage of anticipation occurs about three-quarters through the mission.

Russian researchers Myasnikov and Zamaletdinov report that cosmonauts initially experience excitement characterized by high vigilance and accentuated feelings of personal responsibility.[37] (One cosmonaut reported being "alert" while sleeping.) These reactions are particularly intense during critical flight periods such as at launch, docking, and reentry. Only rarely does this excitement affect performance, and crewmembers experience less excitement over successive missions.

On longer flights, excitement gives way to exhaustion and aesthenia, a nervous or mental weakness symptomatized by sensitivity, irritability, and declining motivation. Cosmonauts become sensitive to provocation, speak in a monotone, and give brusque, stock answers to questions. They have increasing difficulty with sleeping and go to bed later and later. To minimize the problems associated with aesthenia, mission managers schedule the most stressful tasks early in the day and encourage low-key if not calming activities prior to bedtime.

In some cases aesthenia gives rise to euphoria. Lasting from a week to a month, euphoria reflects the novel experience of space and a realization that being there is a tremendous personal accomplishment. Euphoric cosmonauts are characterized by exaggerated levels of activity, an underestimation of the dangers and difficulties of the mission, and an overestimation of how well everything is going. They tend to be somewhat inattentive to mission control, but when they do communicate with support crews their communications are lengthy, emotional, and dwell on grandiose topics. Support crews respond with a professional, task-oriented demeanor, more stringent monitoring, and continued enforcement of the schedule.

Sometimes these euphoric states are followed by depressed states. If this occurs, cosmonauts tend to evaluate things negatively and become irritable and critical. They have low energy levels and their speech is hesitant and listless. Very little seems appealing or funny. Support crews stress the importance of the cosmonaut's work, praising him or her when it is of good quality. They try to cater to the depressed cosmonauts' individual preferences; for example, by arranging radio conversations with specific people on Earth.

Some cosmonauts develop neurotic states that can be continual, recurrent, or episodic. Symptoms are quite varied and include hostility toward others, defensiveness, diminished initiative in communication, resistance to mission control's attempts to change the schedule, and obsessing about other people's reactions, the mission, and personal plans for the future. Sometimes, toward the end of a long mission, cosmonauts show an accentuation of negative personality traits. This includes egocentrism and selfishness, a very subjective perspective, strange ideas, and occasional impulsive behaviors.

MANAGING STRESS

A proper and comprehensive program places heavy emphasis on prevention and seeks to minimize stress through carefully selecting and training spacefarers (see chapter 6), designing environments to minimize stresses (see chapters 4 and 5), and encouraging harmonious, mutually supportive social relations (see chapter 8). A comprehensive program must provide support for the spacefarers' families and provide good follow-up after the spacefarers return from their flights.[38]

Personal Coping

People have the ability to "cope," that is, deal with difficult situations. Several lines of defense keep potentially threatening conditions from giving rise to overwhelming fear and panicky, ineffective behavior. This is clear in Richard Lazarus's "cognitive coping paradigm."[39] According to this theory, we keep a watchful eye out for threats to our welfare. Our reactions to dangerous conditions are moderated by appraisal; that is, the cognitive or intellectual processes that we use to size up situations and their personal relevance. Primary appraisal refers to determining if a situation is actually threatening or dangerous. Once something (such as smoke in a space station) grabs our attention, we ask ourselves, "Is this relevant for me?" and, if so, "How?"

If in the process of primary appraisal the spacefarer determines that the smoke is harmless condensation, that is one matter. If the smoke betokens a catastrophic failure of a life support system, that is another matter and the spacefarer moves on to secondary appraisal. This involves determining if there are resources available to solve the problem. For example, if it turns out that "where there's smoke there's fire," are fire extinguishers available? Is the fire located in a remote module that can be sealed-off from the rest of the space station? Is an escape vehicle handy? Recognition of harmful or threatening conditions, coupled with little confidence in one's power to negate the threat, leads to fear: a rapid heartbeat, shortness of breath, weakness, trembling, and an almost overwhelming desire to be somewhere else.

If we can't solve the problem—extinguish the fire, repair the damage, escape from the situation—we resort to the next line of defense. We try to make ourselves feel better. In this task we are aided by psychological defense mechanisms that operate at the unconscious level and keep us from being overwhelmed by emotion. Defense mechanisms protect us not only against threats from the external environment but also threats from within ourselves. Because the world is not always safe, because we cannot always achieve the high standards that we set for ourselves, and because we are not always as highly regarded by other people as we might like, mental gymnastics ease us through rough times and make life more pleasant.

Hugo O. Leimann Patt has separated the defense mechanisms popular among aviators (and quite possibly spacefarers) into two sets: a healthy set that helps people weather dangerous conditions, and an unhealthy set whose application can make the situation even more dangerous.[40] Healthy defense mechanisms include denial, rationalization, and identification. Denial is the act of refusing to recognize dangerous conditions and perhaps imagining more favorable circumstances in their place. Pilot Dick Rutan stated that after a propeller spun away from his experimental aircraft, his first reaction was one of denial—"This can't be happening!"[41] A spacefarer might believe that an emergency light went on because of a problem with the sensor, not recognizing that it was due to a rupture in the fuel tank or that, due to a miscalculation, the spaceship is running out of propellant earlier than expected.

Rationalization is the act of concocting a reassuring explanation for something that would otherwise provoke fear or anxiety. Among aviators, this sometimes involves mentally reciting statistics that flying is safer than driving in order to divert attention from a dangerous landing. Rationalizations often take the form of excuses for a personal failure. Even as a person who loses a tennis match might blame the defeat on a sore elbow, the spacefarer who fails to complete an experiment might blame the failure on broken equipment or vague instructions rather than a low level of skill or flagging motivation.

Identification is a matter of imagining that one is similar to a stronger, smarter, or more accomplished person. By adopting the attitudes and mannerisms of someone who is at the pinnacle of success, we strengthen ourselves, at least psychologically. Identification is revealed in such comments as "This is the way that Yuri Gagarin would have handled this situation."

The less healthy defense mechanisms include reaction formation, evasion, and displacement. Reaction formation is the rejection of an unacceptable feeling or emotion by stressing its opposite. The person who is terrified, for example, might hide this from himself in a display of overconfidence and bravado. A person who hides fears by taking big chances is unwelcome because he or she puts other crewmembers at risk. Such counterphobic activities may give rise to an accident-prone personality.

Evasion is psychological paralysis, or the refusal to respond to a dangerous situation. Faced with difficult situations, some people procrastinate or decide that other matters are more pressing. By the time they finally get around to doing something, it may be too late.

Displacement involves redirecting negative emotion to a less threatening target. Rather than having a confrontation with a pilot astronaut, a mission specialist might get in an argument with a payload specialist. Although it seems safer to redirect anger from the high-status pilot to the low-status payload specialist, the underlying issues are unresolved and the payload specialist becomes a scapegoat. Thus, while all of the defense mechanisms take the edge off fright, reaction formation, evasion, and displacement could be accompanied by significant performance problems.

Problem-centered coping strategies are aimed at controlling the situations or events that give rise to threats. Emotion-centered coping strategies are aimed at controlling the emotional response to the threat. Retreating to an unoccupied module or laboratory to reduce the stress of crowding would represent a problem-centered method of coping, while contemplation or meditation would represent an emotion-centered method. Problem-centered strategies make the world less stressful, while emotion-centered strategies make us feel better about the conditions we cannot control.[42]

Skill training, which involves "hardening" specific behaviors so that they will be performed even under dangerous conditions, plays a critical role in managing stress.[43] Knowing exactly what to do in a dangerous situation reduces fear. When a fire occurs, a person who has undergone many fire drills will be less frightened and more likely to respond properly than someone who has no idea what to do. For this reason, people who are expected to do dangerous work are very highly trained. Soldiers disassemble and reassemble their weapons again and again so that they can do so almost automatically under combat conditions. Submariners rehearse the procedures to get to the surface if the submarine fails to rise. Pilots and spacefarers are trained to respond properly to all sorts of emergencies. Being prepared—knowing what one can do to eliminate the threat, to make the situation safe—confers a sense of mastery or control that can itself help relieve stress.

Stress inoculation training (SIT) is based on the assumption that the best methods for alleviating stress depend upon the particulars of the situation and the people.[44] As discussed by Judith Orasanu, this involves three phases.[45] The first phase is education—explaining the nature of the stress and describing what to do about it. The second phase is rehearsal or practice, and the third phase is application. For example, a person who is fearful of public speaking might learn a number of techniques for controlling this fear (memorizing opening lines, applying special breathing techniques). Then he or she might practice in front of a mirror and, finally, onstage. Depending on the specific pattern of techniques involved, SIT bolsters both threat-centered and emotion-centered coping skills.

Peer Support

Fellow crewmembers play important roles in helping one another weather stress. One of the most powerful motivators in stressful situations is the desire not to let down our closest associates. Studies from World War II on suggest that, in combat, soldiers take big risks not because of patriotism, orders, or innate aggression, but because they want to help the members of their immediate combat team—in other words, their buddies.[46] In Antarctica, too, people take group identity seriously and fight stress to remain a fully contributing member of the team.[47]

Desmond Lugg raises the interesting possibility that the culture of isolated and confined groups increases the group's hold over the individual and helps reduce stress.[48] For example, a sense of membership in the wintering-over party enhances group identity by underscoring that one is better than the "summer tourists." Some of the behaviors in Antarctica that seem strange may in fact represent healthy coping strategies. Bizarre rituals such as watching a movie over and over, affecting lisps, and using quaint language also can raise group solidarity.

Direct support from other people tends to buffer the effects of stress. Fellow crewmembers will be able to model or demonstrate proper and effective ways of responding and give useful suggestions on how to reduce the stress or at least minimize the aftermath. Peers are sounding boards that help spacefarers better understand their own feelings, as well as guides who nurture them through tough times. They are likely to be excellent sounding boards because they are in the same situation and share the same emotions, something that would not be true of psychotherapists based on Earth. A thorough understanding of the realities of the situation coupled with prior training in psychological first aid constitutes a strong basis for crewmembers' efforts to help one another.

In Antarctica, the community has many ways of handling dysfunction.[49] Sometimes another crewmember is given a specific but not explicit assignment to look after the dysfunctional person and provide for his or her safety. Usually, the dysfunctional person's work is taken over (especially if it is critical) and, if possible, he or she is given a less demanding task. Typically, this is done in such a way as to minimize publicity and preserve the person's sense of self-worth and dignity. It may be done so informally that even the station manager may not be aware that it is happening.

Psychological Support Groups

Russians report success using psychological support groups.[50] For the Russians, psychological support is an ongoing process that begins prior to the mission and continues until long after the flight is over. Psychologists monitor the cosmonauts in flight, keep careful track of their symptoms, and find ways to help them weather the experience.

Ground-based analysis of voice communications provides the major mechanism for appraising the psychological welfare of the crew. The Russians use special filters to identify harmonics or overtones suggestive of stress, as well as analyzing volume and rate of speech, intonations, pauses, and other cues to emotional states. Ground-based personnel who talk with the cosmonauts in space are taught to recognize normal speech and the signs of tension from work overload, fatigue, and emotion. The suspected signs of tension are then verified by subjecting tape-recorded segments of the cosmonaut's speech to computerized voice analysis (a technique that has not yet been proven effective). When video is available, the psychological support team is very attentive to the cosmonauts' nonverbal communication, or body language.

Members of the support team offer cosmonauts a sympathetic ear and counseling. Both spacefarers and their families are given advice on how to deal with troubling issues. If undue stress is identified, the support group takes corrective measures. For example, work schedules might be rearranged to allow the cosmonaut some time for rest and recuperation, and temporarily sagging moods might be uplifted by piping in lively, dynamic music.

The support group also sustains morale by broadcasting news and various sounds from Earth and by arranging for cosmonauts to talk via radio with government leaders, entertainment personalities, and other prominent figures, as well as have private conversations with their families. In addition, resupply vehicles that arrive at Russian space stations carry such items as videotapes and cassette recordings, mail from home, fresh foods, and other treats.

During the 1990s, NASA began psychological support activities under the psychologist Al Holland of Johnson Space Center. The program is similar in many ways to the Russian program, and the kinds of support services provided are based on in-depth understanding of each astronaut's personality and needs. It appears that Dr. Holland's efforts are very welcome.

Psychiatric Health Maintenance Facilities

Because serious behavioral problems will at some point appear in space, there is strong justification for including a psychiatric component in the health care program for extended-duration mission facilities.[51] The goal is to provide on-site management of the problem so that it does not force early termination of the mission or necessitate an expensive or dangerous evacuation.

Provisions should be made to diagnose and treat the disorder. An accurate diagnosis is important because it may reveal an unsuspected problem onboard (for example, the presence of toxic fumes that produce brain injuries with behavioral consequences), as well as because diagnosis is an essential prerequisite for prognosis and treatment. Restraints should be available in case someone becomes violent or starts behaving in ways that could jeopardize the entire mission, but these would be used rarely if ever.

Treatment may be drug based, talk based, or some combination. Recent decades have seen phenomenal advances in psychopharmacology; that is, the development of psychoactive drugs that tend to minimize, even eliminate, behavioral and other symptoms of mental disorders. Different classes of drugs are intended to control disorganized thinking, reduce anxiety and phobias, lift people out of depression, and so forth.

A good sampling of these drugs may be carried on the ISS. Unfortunately, they are not as simple to administer as decongestants or aspirins. Oftentimes, their effectiveness depends on highly accurate dosage levels (which may depend in turn on blood analysis or other unavailable laboratory work), and their effects on their users may not stabilize for a long time.

Talk therapy, notes Santy, must be brief, focused, and to the point. The whole idea is to ease the person through the troublesome situation. In-depth therapy for problems dating to early childhood will have to wait. One possibility is that fellow crewmembers will have enough training to provide emergency counseling or therapy for certain problems. There are forms of counseling that can be applied by people who are not highly trained mental health professionals but who are willing to serve as peer counselors or skilled helpers.[52] Or perhaps someday psychotherapy will be conducted by video or e-mail by a psychotherapist stationed on Earth.

CONCLUSION

In the late 1950s, notes Peter Suedfeld, as humans prepared for their first forays into space, we had relatively little data to draw upon in our efforts to forecast psychological and social adaptation to spaceflight.[53] Much of the available evidence came from early polar explorers and Operation Deepfreeze, which deployed large numbers of military personnel to Antarctica. The reports of earlier explorers included vivid descriptions of hardship and deprivation and the sometimes terrible toll it took on the men. At about the same time, studies of sensory deprivation, which involved subjecting people to conditions so that they would receive as little stimulation as possible, were finding that people became confused and disorganized, highly emotional, and subject to hallucinations.[54] This led to dire predictions concerning life in outer space.

If we were preparing to enter the space age today we would be more optimistic. Over the past forty years, as we have accumulated more and more information from polar bases, research submersibles, nuclear submarines, and space itself, we have discovered that people's reactions are not as severe as we expected based on early reports. Perhaps some of the early explorers exaggerated their experiences for dramatic effect, perhaps not. What is clear is that in comparison to the conditions encountered by early explorers, living conditions are much more benign today. Furthermore, we are better equipped to screen out people who are ill-suited for remote duty, and transportation and communication technologies provide safety nets that the early explorers would find hard to imagine.

However, Suedfeld adds that because the initial extreme predictions were not confirmed, we may have become too complacent about people's abilities to withstand the stresses of spaceflight. Furthermore, we must be aware of the possibility that stressful episodes are not reported or are downplayed. Succumbing to stress, particularly by men, is devalued in our culture. As Santy points out, some of NASA's actions suggest that the agency used the psychological process of denial to wish away the potential difficulties attendant upon psychological problems.[55] NASA may have taken steps to cover up certain episodes, and it has certainly discouraged qualified researchers as well as snoopy reporters.

Individual spacefarers are assigned relatively few flights, and any hint of weakness could disqualify one from participation. NASA's reaction to any kind of medical problem, says Deke Slayton, has been simply to send someone else, even if it was virtually certain that the problem would have no effect whatsoever on the success of the mission.[56] The fear of risking the wrath of unhappy NASA managers, and the fear of embarrassing oneself in public, encourages spacefarers to hide any signs of doubt or tension. If both managers and astronauts cover up psychological issues, the end result is unrealistic perceptions of life in space and an inability to prevent such problems or to cope with them properly when they do arise.

Spacefarers themselves have no monopoly on stress. NASA officials, contractors, flight controllers, flight surgeons, family and friends, and even media representatives have fallen prey. For example, television networks gave the space program extensive coverage during the first decade of spaceflight. Apart from the actual launches (which could be repeatedly delayed), there was not very much real action to televise. Nonetheless, notes Howard E. McCurdy, commentators enthralled audiences by talking with NASA officials and the contractors who developed the rockets and space capsules, by showing off space suits and other pieces of equipment, and by explaining the scientific aspects of the project.[57] Jules Bergman was among the newscasters who became media celebrities by explaining technology using simulators and props. During Scott Carpenter's Mercury flight, Bergman himself donned one of the medical harnesses that astronauts wore so that ground personnel could monitor vital signs. This device showed that "Bergman suffered as much stress during the twelve hours of television coverage as Carpenter did in orbit."

CHAPTER 8

GROUP DYNAMICS

As he prepared for one of the last Apollo Moon flights, a primary goal of Eugene Cernan was to weld three people into a crew.[1] As commander he knew he had to consider "more than switches and numbers." In his estimation, "people problems usually are much more difficult to handle than mechanical glitches," and "any failure on their part would be my failure too." To succeed as commander he had to understand his mates on a personal level, know what made them tick, and earn their confidence in him as the mission's skipper. His views are fully consistent with modern conceptions of leadership, which state that not only must leaders have a good understanding of the technical aspects of their work but they must understand the interpersonal dimensions as well.

Space crews are teams: groups of two or more interacting people who have a stable relationship, share common goals, and see themselves as forming a team. These team members have task-relevant knowledge, carefully defined roles and responsibilities, and complementary skills. They are highly interdependent in that they need one another to survive and get the job done, and they achieve very high levels of coordination in the process. Team members are bound together by common interests, interlocking skills, and, we hope, friendship. Later on, spacefarers may exist who do not qualify as members of teams. Consider the independent scientist, just hopping a ride to a destination to do independent research, or members of a maintenance crew working side by side and all doing the same thing. At large extraterrestrial bases we may find people who do diverse, unrelated types of work.

CREW COMPOSITION

Crew composition refers to the number and kinds of people who are included in the crew. Over the years, crews have become larger for the simple reason that completing today's complex missions requires higher levels of staffing than were required in the past. Crews are becoming more heterogeneous or diverse, in part because we are overcoming the expediencies, fears, and prejudices that restricted the initial pool to military test pilots, and in part because candidate pools remain unnecessarily limited if we needlessly exclude broad segments of the population. As we undertake new types of missions we will have to draw on a greater range of people to assemble the necessary skills.

Crew Size

Because of the high cost of sending people into space, it is tempting to keep crew sizes as small as possible. Not only must we launch each additional astronaut, we must also launch the extra food, water, living space, and everything else needed to support that person in space. Space stations have typically housed 3 to 6 spacefarers (occasionally more, when a shuttle is docked at Mir), and at maturity the ISS is likely to have a crew of 6. There are varying scenarios for the initial Mars crew, ranging from 4 members under the more economical scenarios to 8 or so in the more lavish. The shuttle can carry as many as 10 people, but mostly on one- or two-week missions. Crews in the small group range (roughly, no more than a dozen members) will be standard for a while, but size will increase again if we establish a permanent presence on the Moon or develop a community on Mars.

Larger crews bring a greater range of abilities, talents, and skills to the enterprise and have greater backup capabilities in the event that someone is incapacitated. Larger crews offer the additional advantage of greater social variety. Having more crewmembers or socializing with different crews means an expanded range of friendship choices and more relief from monotony or boredom.

Laboratory studies suggest that isolated 3-person groups get along better than their 2-person counterparts, unless the group is crammed into a tiny interior space.[2] In a twenty-one-day simulation study, 2-person and 3-person groups were confined in settings that allowed either 60 cubic meters per person or 21 cubic meters per person. Under the less-crowded conditions, the 3-person groups reported lower stress and anxiety and complained less as the study progressed. The extra person enriched the social environment, but this advantage was lost in the more crowded conditions.

In Antarctica, social relations are smoother in relatively large groups, but this finding is difficult to interpret because small groups tend to be housed in primitive facilities, whereas large groups tend to have better quarters. One study found that groups of 8–10 members were less compatible and more hostile than groups of 20–30 members, and another found greater emotional and interpersonal problems at stations of 15–40 members than at even larger bases.[3]

Kanas and Fedderson propose odd-numbered rather than even-numbered crew sizes because even-numbered size crews would be too likely to split down the middle when making a decision.[4] For example, a 4-person crew bound for Mars might vote 2-2 and reach an impasse, but a 5-person crew would have to vote 3-2 (or better), thus preventing a deadlock. However, it might not be so simple. On arrival at Mars, an odd-numbered crew might have to split into two subgroups (one to orbit the planet and the other to explore the surface) and one of these two subgroups will necessarily be even-numbered. Furthermore, having an odd-numbered crew is not an imperative since we can impose rules designed to prevent deadlocks—for example, allowing the mission commander to break a tied vote.

Age

Given that a person is fit to withstand the spaceflight environment and do the work, age may confer an advantage, particularly for people in leadership positions. An early study of residents of an undersea research habitat found that mature group members were "parent surrogates" to younger aquanauts.[5] This does not necessarily mean that the most senior member of the crew is the official leader. Russian spaceship commanders, chosen from the ranks of the military, are sometimes ten to fifteen years younger and much less experienced in space than their civilian flight engineers.[6]

Astronauts are rarely chosen in their twenties and may not fly until their late thirties or beyond. (This was particularly true for astronauts selected at the end of the Apollo era who then had to wait several years for the shuttle to become operational.) Some of the most successful and respected spacefarers continued to fly into their sixties. It is tremendously expensive to develop seasoned spacefarers, so arbitrary age restrictions work against getting the best return on our investment in training.

Until late 1998, the astronaut Storey Musgrave held the age record, as he had flown in space at sixty-one. He lost this record when NASA sent John Glenn aloft at age seventy-seven. Glenn, who was the first astronaut to orbit Earth, on February 20, 1962, had maintained his airplane pilot's license and passed the physical. He had been grounded after his first flight (as was the first cosmonaut Yuri Gagarin) because the U.S. was afraid of the devastating effect on morale if this space hero were lost during a subsequent flight. In 1998 Glenn was assigned to a shuttle mission. Whereas some people considered this a "victory lap," his official purpose was to conduct research on aging and the biomedical consequences of life in space.[7]

Gender and Ethnicity

In the early 1960s, the Russians launched the first woman, Valentina Tereshkova, into space. She then left the cosmonaut corps. Suspicions that her flight was largely for propaganda purposes are supported by the fact that many years would pass before women would again fly in Russian craft and that, with the exception of Svetlana Savitskaya, who flew on Salyut 7 in 1982 and 1984, and Yelena Kondakova, who flew on Mir in 1994, the women who did fly in Russian craft were English, French, or American rather than fellow Russians.

The world of 1959, when the first astronauts were chosen, was quite different from today's world. Many people at that time had stereotyped ideas about the proper roles for men and women, and with some exceptions, neither men nor women were as aware of women's potentials as they are today. Scant attention was paid to the famous women pilots of the 1920s and 1930s or to the women who tested aircraft during World War II and in the process logged thousands of hours of flight time ferrying war planes from one destination to another.

After the original Mercury astronauts had been selected, a wealthy and politically savvy aviator, Jackie Cochran, arranged for several experienced women pilots to undergo physicals at the Lovelace Clinic, where the same tests had been conducted on male candidates.[8] These women were well on their way to proving their qualifications, but at a critical time Cochran withdrew her support, believing that diverting attention from the men who had already been selected could force the

United States to fall further behind the Russians. As a result of this mixture of events, she is viewed by some people as a pioneer who helped women into space, and by others as someone who systematically blocked women's progress through the Apollo program.

Nonetheless, by the mid-1980s several women astronauts were in training. Sally Ride was the first American woman to fly, and she did so as a mission specialist in 1983. Other women followed. Eileen Collins became the first woman shuttle pilot in 1995 and the first woman shuttle commander in 1999. Shannon Lucid spent several months on Mir. As of 1996, women comprised 18 percent of the astronaut corps. A series of interviews led Cynthia Griffin to conclude that, despite varied backgrounds, interests, and personalities, they are all "highly trained professionals and are serious about their life-threatening work." She added that they feel "comfortable and equal working along side their male colleagues."[9]

Similarly, black Americans and other ethnic minorities were slow to appear in the astronaut corps. Very few were selected, and some people suspected that the initial selections were politically motivated.[10] Yet here, too, diversity is on the rise. Shuttle crews have included Asian Americans, black Americans, and Hispanics, but as in other occupational settings, diversification progresses at a very slow rate.

People's perceptions and reactions, rather than gender-related or race-related performance problems, constitute the major barrier to increasing the diversity of space crews. Since, on the whole, women are smaller, require fewer calories, and are more resistant to problems associated with crowding, they may be *better* suited for spaceflight than men.[11] Yet the same immaturity, prejudicial attitudes, and other factors that slow integration into other work settings may work against newcomers to space. The first black, for example, who enters a formerly all-white setting operates under tremendous pressure. She has to prove herself in ways that her colleagues do not, and her coworkers are likely to be skeptical if not hostile. Mistakes or failures barely noticed if made by a white male may be magnified and attributed not to personal error but to gender or race.

Envisioning future space settlements, T. A. Heppenheimer argues that the workforce should have approximately equal numbers of men and women.[12] In his view, slovenly behavior, aggressiveness, and other problems are associated with prisons, army camps, boomtowns and other all-male or predominantly male communities. A more balanced population would allow people to meet, fall in love, start families, and in other ways establish lives that resemble those that many of us on Earth enjoy.

International Crews

Space exploration and settlement are international ventures, if for no other reason than because nations join forces to finance such expensive undertakings. At this point, we have had considerable experience with international crews. This includes the Apollo-Soyuz mission of 1975, the Russian "guest cosmonaut" program, which began in 1978, the inclusion of international crew members on shuttle flights since 1983, and astronaut flights on Mir.

A survey of fifty-four spacefarers suggests that the greatest challenge is acquiring foreign-language proficiency—not simply learning minimal skills but knowing the language well enough to communicate clearly, even under difficult conditions such as on the radio (where there are no nonverbal cues to reinforce a message) or during an emergency.[13] European Space Agency researchers estimate that ESA astronauts require three hundred hours of basic training and two hundred hours of maintenance as they prepare for a Russian mission and develop not only proficiency in basic Russian but also a good understanding of Russian technical terms and "space slang."[14]

In his fascinating book *Dragonfly: NASA and the Crisis aboard Mir,* the journalist Bryan Burrough describes problems encountered by U.S. astronauts on Mir.[15] He attributes strained interpersonal relations and a sense of social isolation to certain astronauts' personalities. Astronauts who fared better, he suggests, were those whose personal traits allowed them to get along with the Russians.

Language fluency, pre-mission crew training, and ease of communication with Earth also affected astronauts' adaptation to life aboard Mir.[16] Norm Thagard, the first to fly, felt isolated by the Russian language and culture. Andrew Thomas, the final Mir astronaut, was assigned to his crew very late and suffered because he did not have enough time to practice Russian. John Blaha was fluent in Russian, yet he felt isolated since the crew to which he had been assigned was replaced at the last minute. Early Mir astronauts had limited contact with the ground, making it a common situation for Thagard, Lucid, and Blaha to go for days at a time without hearing the English language on the radio. Frustrated by the poor voice link, Linenger began to rely on e-mail. For those who came after Linenger, e-mail became a fair substitute for conversation.

Another challenge is bridging cultural differences in such divergent areas as dietary preferences, recreational interests, and privacy. Special arrangements may be necessary to accommodate religious practices, as they were on a shuttle flight that carried a Saudi Arabian prince. And of course, it may be necessary to overcome prejudices based not only on race or ethnicity but also on politics; after all, the Cold War is not that far behind us. People can openly disavow prejudices but still allow old suspicions to affect their thinking.

According to Harland, there were practical as well as political reasons for bringing visitors to Salyut and Mir.[17] In the late 1970s the Russians estimated that the Soyuz ferries docked at the stations were able to withstand the harsh orbital environment for about ninety days. Since crews would stay aboard for half a year or more, the ferries had to be rotated, and in the course of this rotation guests could be brought aboard. The guest cosmonauts' primary assignments were studying their home countries from space. Most guests had extremely fast-paced schedules—sleeping as little as three hours a night—and returned to Earth totally exhausted. Sometimes the hosts' performance deteriorated as they spent time on ceremonial activities and showed the guest around. The guest cosmonauts' training was not as extensive as that of the rest of the crew, so they had to be watched carefully. It was not always possible to regulate the guest cosmonauts' schedules.[18] In 1987, one crew was surprised to discover that a previously scheduled "visiting" crewmember would stay aboard to replace the long-term crewmember whose EKG readings worried the doctors.

U.S. shuttle missions have included international crewmembers (ICMs) from Canada, France, Germany, Holland, Japan, Mexico, and Saudi Arabia. A debriefing of ten U.S. astronauts who had flown with an ICM identified 9 preflight, 26 in-flight, and 7 postflight instances of miscommunication, misunderstanding, or interpersonal conflict.[19] Findings confirmed earlier reports that ICMs receive less than adequate training. U.S. astronauts noted potential conflicts due to differences in personal hygiene, housekeeping practices, and attitudes toward the mission.

How can we reduce the potential for international misunderstanding? Cultural sensitivity could be a basis for choosing people for international missions. Education and travel that sensitize crewmembers to different cultural traditions and practices may help them become more accepting of one another during a mission. ESA language training includes materials intended to help European spacefarers become familiar with Russian culture. Training periods of six months or less are not long enough. Failure to begin integrating the ICM into the crew during the crew's formative stage may intensify "in group" and "out group" distinctions. On the whole, life is not easy for newcomers entering established groups.

Through training, people develop mutual understanding, build bonds, and work through difficulties *before* they are sent into space. Such training should include a psychological component that has both didactic and experiential aspects. Another remedy is to ensure that the entire crew has good times together prior to departure. In preparation for the Apollo-Soyuz mission, U.S. astronauts hosted Russian cosmonauts on a hunt in Montana; the astronauts made arrangements so that the two groups would leave their nonastronaut managers and escorts behind.[20]

GROUP STRUCTURE AND PROCESS

Group structure defines the relationships among different members of the group. It prescribes their social roles (rights, obligations, and ways of behaving) and determines their social status, that is, the esteem in which they are held by others. Social structure is evidenced in organizational charts, policies, and regulations; official job descriptions; and informal rules regarding how things should be done. Group processes such as communication and decision making are channeled by structure and revealed in the ebb and flow of social interaction over time.

Authority refers to the power vested in a position. People who are high in the pecking order (the NASA administrator, the chief astronaut, the mission commander) have the right to tell underlings what to do. People who outrank us may have additional forms of power, such as the power to reward us (put us in line for choice assignments) or punish us (assign us unpleasant duties or get us fired). They may know important inside information that we would like to have, such as knowledge of upcoming assignments. By selectively withholding and divulging such information, they can exert even more control.

The power of social roles is evident in one of the tragedies that befell mountain climbers ascending Mount Everest in the spring of 1996.[21] On that ascent there was a sharp distinction between the guides, who had absolute authority and were not to be questioned, and the clients, who had paid fees of up to sixty-five thousand dollars to be escorted to the summit. The clients were quite varied in terms of fitness and ability, ranging from those with extensive climbing records who had already ascended the mountain (and who found the price cheap relative to organizing their own expeditions) to marginally competent but wealthy people intent on realizing their personal dreams.

One client was a highly fit journalist. As the members of this particular expedition were straggling down the mountain during a ferocious storm, a guide approached the journalist client to see how he was doing. Of the pair, the guide was suffering the most from exertion, lack of oxygen, freezing temperatures, and other rigors of the climb. In true superior-subordinate fashion, the conversation dwelt on the welfare of the journalist. Yet, it was the guide who died on the mountain and the journalist who returned home. In retrospect, the journalist realized that the guide's fatigue and confusion were apparent, but the structure of the relationship was such that the issue of the guide's welfare never came up. If the guide and the journalist had been participating as equals, the journalist believes, he would have immediately recognized the true condition of the guide and insisted that they both return to camp, instead of standing by while the guide wandered off looking for stragglers.

Leadership

Leaders take the initiative in social situations, plan and organize action, motivate others, and elicit cooperation to reach common goals. In the course of this, leaders help the group understand the task at hand, work cooperatively and harmoniously with one another, and withstand threats from outsiders. It's tempting to think of leadership as a one-way process with the leader telling everyone else what to do, but a closer look reveals that leadership is a much more complex interactive process. Leaders respond to suggestions, advice, encouragement, and criticism. In addition, by recognizing and accepting a person as leader, group members validate the leader's status.

When we interpret other people's behavior, we are more likely to focus on their skill, motivation, personality, and other internal forces than on the environmental conditions that act upon them. For example, we may be more likely to attribute aircraft accidents to pilot error than to misleading or faulty information presented to the pilot in the cockpit. This focus is particularly sharp in Western societies, which emphasize individual responsibility and initiative. Given this focus, it is not surprising that many discussions of space crew leaders tended to dwell on leaders' traits.

Jack Stuster's review of effective leaders in isolated and confined settings suggest that, among other things, leaders are alert and have good problem-solving abilities, have good interpersonal skills and the motivation to apply them, are democratically oriented, have high self-confidence, and are credible, flexible, and adaptable.[22] Joseph Kubis describes the effective space crew commander as competent, goal or achievement oriented, and interpersonally sensitive, with an awareness of human needs and the importance of opportunities for their satisfaction.[23] John Nicholas and Larry Penwell suggest that successful extended-duration crew leaders will be hard working, optimistic, and respectful of the crew. They should be able to take charge during emergencies but otherwise follow a more democratic approach that allows crewmembers to participate in the decision-making process.[24]

As implied by these descriptions, leaders must engage in two types of activities: instrumental activities that get the work done, and relational activities that promote satisfaction and harmony within the group. Instrumental activities are emotionally neutral acts that include orienting group members to the work, providing fuel for the discussion, answering their technical questions, organizing and coordinating them, and giving them a sense of purpose and direction. Relational activities are emotionally toned acts aimed toward maintaining positive attitudes, preserving group integrity, and building morale. Giving people emotional support by listening carefully and offering reassurance, soothing hurt feelings, and telling jokes to break the tension are among the relational activities that help build morale.

According to Robert Helmreich, Clay Foushee, and their associates, the strong, silent, macho or "stick and rudder" individuals who keep their emotions to themselves and lack sensitivity to other people may have made great test pilots and fine choices for solo space missions.[25] However, when we turn to crews consisting of two, three, or more individuals—such as crews of large, multiengined aircraft or space missions from Gemini forward—relational skills are equally important.

One early study found that, in comparison to less effective airline pilots, more effective pilots recognized the need for sensitivity to other crewmembers' feelings.[26] They knew that it was important to communicate with other crewmembers so that they all had the necessary information. These pilots did not see themselves as infallible, and they encouraged subordinates to question their actions when necessary. The less effective pilots tended toward more authoritarian or directive leadership styles and were less likely to acknowledge their own limitations.

On the whole, pilots of multiengined aircraft who are sensitive to other people are rated more highly by their crews and perform at a higher level than pilots who have high technical competence, limited interpersonal skills, and a large ego. It is leaders who have high scores in both dimensions whom we hope to find leading space expeditions.

Perhaps instrumental and relational activities are best viewed as tools that can be used as needed. If crewmembers are uncertain what to do, the leader may have to draw heavily on instrumental skills. If, on the other hand, there is flagging motivation or conflicts that keep the group from moving ahead, the leader may draw on relational skills to resolve conflicts or encourage buy in. Then again, if the crew is both willing and able to do the job, perhaps the wisest strategy for the leader may be one of receding into the background and letting the group forge ahead.[27]

Communication

Prompt, accurate communication is vital. If crewmembers remain silent at the wrong time, their fellow crewmembers may not receive essential information. Yet, if everyone continually bombards everyone else with information, progress slows and chaos could reign. If there is an imbalance it is usually due to too little communication rather than too much. For example, if an aircraft or shuttle pilot seems deeply preoccupied or somehow unapproachable, subordinates may fail to volunteer crucial information. In fact, some aircraft accidents might have been prevented if there had been more open communication with the pilot.[28]

Mary M. Connors points out that several factors complicate person-to-person communication in space.[29] The high noise levels during takeoff can make it difficult for people to communicate with one another. The high background noise emanating from life support equipment can make it difficult for people to hear one another, particularly if there is reduced air pressure within the cabin. Astronauts reported that, in the cavernous interior of Skylab, their voices dropped off quickly and they had to shout to be heard. International crewmembers who lack fluency in a common tongue may have difficulty following instructions (particularly in dangerous situations when they are under stress), and they may run afoul of interpersonal difficulties. A crewmember who inadvertently chooses the wrong words when trying to provide someone from another country with emotional support may come across as sarcastic or hostile.

Much communication occurs nonverbally—through facial expressions, postures, gestures, and tenor of voice. Unfortunately, under conditions of microgravity fluids tend to pool in such a way that spacefarers' faces seem puffy and less expressive. In addition, as a spacefarer floats around in various positions and orientations, his or her body language may be lost. Furthermore, notes Connors, distancing cues become less reliable. One way we can tell how other people feel about us is by how close they stand. On Earth, it takes energy for people to move away from one another, but in microgravity it takes energy for them to remain nearby.

Alan Kelly and Nick Kanas's survey of spacefarers underscores the need for fluency in a common language and confirms that ambient noise and facial swelling hinders communication between crewmembers.[30] Space sickness also makes it difficult to communicate—perhaps no big surprise. Certain types of complex communication, such as reading, writing, and gesturing, tend to decline over time. On the other hand, shared experiences and the excitement of spaceflight, living together in close quarters, and isolation from Earth tend to facilitate communication among spacefarers.

Conformity

Group norms are the formal and informal rules that prescribe particular ways of looking at the world and doing things. Norms define what is right and proper in the eyes of the group.[31] Norms arise in part as carryovers from other situations. Some of the military norms prevalent in the early days of spaceflight were carried over into the civilian space program. Norms are based also on precedents set over time. Something that is done in the same way again and again becomes "the way things are done around here." Generations of cosmonauts have watched a patriotic movie entitled *The White Sun of the Desert* the night before departure, and have (in an unrelated ritual) urinated on the right rear bus tire at the launch site. (Women take part by bringing bottles of urine.)[32] Norms are based also on specific events in the group history. Certain critical incidents involve behaviors that become standard later on.

People who step out of line in isolated environments may endanger everyone else. Because of this, such settings may intensify pressures to conform. Deviants who violate the ways of the group are likely to provoke sharp comments and other actions intended to get them to shape up and "get with the program." People respond to nonconformists first by directing more comments their way and then by ostracism; that is, by treating that person as if he or she no longer exists. The recalcitrant nonconformist is psychologically if not physically rejected from the group.

Observations of Antarctic personnel underscore two adverse consequences of rejection. First, the rejected person may simply drift off mentally and in this way deprive the crew of his services. Whereas in some Antarctic settings it may be possible for someone else to take up the slack, this is less likely to be true in space where crews are kept as small as possible. Second, because the crew is isolated there is nowhere else for that person to turn for companionship. After a tiff with one's family, coworkers may seem warm and nurturing, and after a dispute at work it is the family that comforts one. In an isolated location there are no such buffers, and a downward spiral may ensue. In Antarctica and perhaps in space rejection can lead to the long-eye syndrome also known as a "twelve mile stare in a six mile room."[33]

Cohesiveness

Cohesiveness refers to the bonds that tie group members to one another. Cohesiveness is, in essence, the social glue that makes some groups more grouplike than others. In comparison to members of noncohesive groups, members of cohesive groups take greater pleasure in group membership, are more involved in the group, and are more dedicated to the group's goals. Unless we are contemplating our enemies, we prefer a highly cohesive group to a loosely knit assemblage of people who care little for one another and lack commitment to the cause.

Most of the conditions known to strengthen cohesiveness should work to the benefit of space crews. People take more pride in reaching difficult goals than in attaining easy ones, and a high price of membership (or severe initiation) tends to fortify a group.[34] There certainly is a high price of admission to a space crew. A spacefarer's initiation includes surviving a long and tough selection process, undergoing rigorous training, enduring dangerous and difficult living conditions, and all of the other challenges discussed in this book.

Groups take pride in seeing themselves as set off from and better than other groups. This is illustrated by the sense of elitism among those Antarctic personnel destined to stay for the winter, who feel superior when they compare themselves to "summer tourists," who will soon depart for home. Expedition astronauts selected for a special mission (such as a trip to Mars) may consider themselves better than crews selected for routine shuttle flights. Most crucial of all, the mission itself is a goal of overriding importance that should maximize teamwork and minimize conflict.

Cohesiveness tends to increase adherence to group norms. If group norms favor high levels of productivity and achievement, as is likely in the case of a space crew, then high cohesiveness is associated with high performance. If, on the other hand, norms favor doing as little as possible or goofing off (such norms sometimes develop among some work groups), then high cohesiveness may actually undermine performance. There seems to be little risk of this right now, but the situation may arise at some future time when space becomes home to large numbers of industrial workers.

Cohesive groups tend to have tight boundaries. There is a sharp distinction between the "in group" and the "out group." This helps the group ward off threats from external parties but may also contribute to intergroup conflict. Tight boundaries also make it difficult for an outsider or newcomer to join a group. This was seen in the FNG (fucking new guy) phenomena among U.S. troops in Vietnam, which kept newcomers from being assimilated into a combat team until they had proven themselves under fire.

Decision Making

We can arrange decision-making processes along an autocratic-democratic continuum. At the autocratic end of the scale, the leader makes unilateral decisions. As we move from the autocratic to the democratic end of the scale, we find a number of strategies whereby the leader consults with followers but then makes the decisions on his or her own, and then strategies where people vote. Democratic decision-making includes representative democracies, where a few people vote as representatives of their constituencies, and full democracies, where everyone votes. Finally, there is discussion to consensus. Under the latter procedure, the group keeps talking about the issue until everyone reaches full agreement.

The best or most effective strategy depends on several considerations.[35] One is the amount of time available. If there are tremendous time pressures, then the leader is not able to solicit the members' opinions and an autocratic decision is the only possibility. The leader's expertise is important too. If he or she is more knowledgeable than the average crewmember, then attending to followers' ideas could lower the quality of the decision. If the leader is relatively inexperienced, it makes sense to listen to the wisdom of the group. Another consideration is group buy in. In part because we like to be consulted, and in part because we have a better understanding of decisions that we have helped make, we are more likely to accept democratic decisions. If the group reaches consensus there are no "losers," and for this reason the decision is easy to implement.

When should decisions be made by support personnel and when should they be made by the commander and crew in space? This is not an easy call. The actual flight crew is only a small part of the mission. It also includes people who track and monitor the flight, undertake recovery, and the like. It would not do, for example, for a crew to make an arbitrary, unilateral decision to remain in orbit for a couple of extra days or come home a week early. Under many conditions mission control makes the difference between life and death. On occasion, fellow astronauts have entered simulators and solved problems that were endangering their colleagues in space. Yet, there are many areas where a decision could be made on the ground or in space.

NASA has a reputation for micromanagement, that is, looking over the astronauts' shoulders and controlling their every move. In the early days of spaceflight, this made considerable sense. Ground-based personnel using telemetered data had a better understanding of the course of the mission than did the flying astronauts. It was ground personnel, not the astronaut, who had evidence that the Mercury capsule's reentry shield might be detached and who struggled with the possibility that John Glenn would be incinerated on reentering the atmosphere. Communication with the space capsule was instant, assured by strategically located centers that provided almost continuous communication as the capsule orbited Earth. Onboard computers of that day were limited by today's standards. Retaining centralized control—trying to manage everything from Earth—makes sense when people in Moscow or Houston have a lot more knowledge and expertise than do people aboard the craft, when they have a good understanding of the conditions confronting the spacefarers, and when there is excellent communication between Earth and space.

Today, powerful computers are carried into space. These contain data banks and programs that will help the crew make good decisions onboard. As crews become larger, there will be more expertise within the crew, and this also will contribute to good decisions at the local level. As spacefarers travel to Mars and beyond, communication with

Earth will take longer and longer. These delays mean that crews must become more self-reliant, particularly in emergency situations that demand prompt responses.

Shifting decision-making authority from NASA to the space crew is consistent with the emerging managerial practice of empowerment; that is, letting the people who are closest to the problem make the decision. Empowerment means that mission planners and managers would disseminate information to crewmembers rather than "playing it close to the vest." They would ask crewmembers questions rather than tell them what to do, give them the freedom to make choices, and provide the support they need for implementation and follow-through. Empowerment is based on the belief that (1) those closest to a problem are the ones most likely to find a solution to it, (2) jobs become more rewarding when people make decisions for themselves, (3) work is done quicker because there is less need for information to be shuffled from place to place, and (4) the manager's time is better spent on other activities such as working on budgets or planning the next mission.

NASA is changing its communication strategy as we enter the era of the ISS, and this will result in less communication with the crew.[36] There will be fewer people in the Mission Control Center, except during some specialized operations. This gives the support personnel freedom to work on other jobs during the long ISS missions. (They can be rapidly summoned if need be.) Eventually, computers will communicate between ground and spacecraft, and people will be out of the loop unless the computer determines that human intervention is necessary. Otherwise operations will proceed with minimal intervention from the ground.

CONFLICT

Interpersonal tensions are heightened as a result of the frustrations and hardships of the environment coupled with no opportunity for escape. People in isolated and confined groups are aware that the situation is explosive, and they work hard to keep their feelings under control. Some believe that they must "walk on eggshells," and they long to return to a place where they can be more forthright and direct when dealing with others. During one early Salyut mission, two cosmonauts had barely settled in when they had a falling out, and they essentially ignored each other for the duration of the mission, even though they were cooped up in a single module.[37] In early 2000, during a long-duration mission simulation for the ISS, two male participants got into a drunken brawl and pressed unwanted affections on a female participant. She successfully fought them off, but for the remainder of the confinement the knives were carefully hidden from the brawlers. The chances of friction with another crewmember may increase with mission length, but it is not necessary for people to be cooped up with one another for a month or more to find one another irritating.

Unpleasant personal characteristics or mannerisms loom large in isolation and confinement. Perhaps because people who are dirty, who snore, or who have other obnoxious habits are impossible to escape, other people's patience with them quickly wears thin. Almost every Antarctic team boasts someone who delights in never bathing and in wearing the same filthy clothes for weeks at a time.

Real or imagined slights may assume large proportions. Chester Pierce described the role of micro-aggressions—barely detectable aggressive acts—that have strong cumulative effects.[38] Micro-aggressive acts include inappropriate facial expressions (such as grimacing or grinning), contradiction, sarcasm, and anything else that undermines the victim's sense of self-worth. Because each of these assaults is small, it is difficult for the aggrieved party to convince others that he or she is being maltreated or to mount an effective defense.

Inequity, or unfair work assignments, is another source of conflict. In a study of a long and dangerous Antarctic trek, Gloria Leon and her colleagues found that giving some people less desirable assignments than others—consigning them, perhaps, to the equivalent of permanent dish-washing duty at home—caused tensions.[39] Perhaps above all, leaders must recognize that part of the job is to listen carefully, offer emotional support, and find creative ways to resolve grievances and disputes without lowering performance expectations or making special concessions. This requires sensitivity, skill, and a determination to meet the relational obligations of leadership.

Factionalism

A classic finding in sociology is that a three-person group will break up into a coalition of two persons against one, "the outsider." A Mir simulation study conducted at Moscow's Institute for Biomedical Problems by Vadim Gushin and his colleagues found that despite the best intentions to form a cohesive group, two different three-person crews split into two-one subgroups, with the majority quite critical of the outsider.[40] The coalition considered the outsider infantile and impractical, and later criticism extended to his professional credentials, as well as his behaviors. These divisions occurred despite efforts by all three persons to view each other positively and to develop cohesiveness by becoming similar to one another psychologically.

There are many ways that crews of different sizes could break up into warring factions. Typically, notes Larry W. Penwell, intergroup conflicts are evidenced in own group bias (an inflated evaluation of one's own group and discrimination against the "out group").[41] Each group tends to deny its own weaknesses and contributions to the problem, and sees the other group as an enemy. Hostility between the two groups increases and communication declines. Leaders adopt more authoritarian styles, and followers become more compliant. In fact, to reassert their hold over their groups, some leaders may agitate their groups about real or imagined enemies.

Some sailing expeditions of the nineteenth century combined scientists and seamen. These two groups, notes Ben Finney, had very different backgrounds and interests.[42] On the whole, the seamen were uneducated and preoccupied by practical matters. They were used to taking their lives in their hands and enduring harsh discipline. The scientists, on the other hand, were educated, oftentimes wealthy, and devoted to abstract, otherworldly pursuits. They were not subjected to the same harsh disciplinary measures as the seamen. When hostility arose between the two groups, the seamen found ways to undermine the scientists' work, either by failing to cooperate fully or by sabotage.

Divisions along occupational lines have been observed in Antarctica and in outer space. Harland reports that in the Russian space program, when a third seat was added to Soyuz and the two cosmonauts were joined by a physician, the doctor was considered an outsider.[43] Similarly, foreign researchers were not viewed as members of the cosmonaut team, but as guests to be looked after. In the U.S. space program, we might expect a division between the pilots and the mission specialists, who are the "real" astronauts, and the payload specialists, who are less versed in spacefaring but have specific scientific or other duties. This division may be exacerbated because the payload specialists tend to be assigned only one mission and may not join the crew until after it has already completed most of its training.

Divisions along gender, ethnic, and national lines are other possibilities. At least one cosmonaut refused to serve on the ISS under an American commander, and was transferred to a Mir crew.[44] Penwell adds that divisions might form among people who live in different modules on a space station or who arrive at a destination such as Mars via a different spacecraft.[45]

Conflicts with Mission Control

Tense relations and conflicts with mission control deserve serious attention. The recent Mir simulation studies conducted by Vadim Gushin and his associates found that, after about a month in isolation, crews tended to become more egocentric and more sensitive when communicating with mission control. Also after about a month, communication decreased. To some extent this is understandable because the simulated crew had many of their questions answered during the early weeks of confinement and developed increasing autonomy over time. However, there was also evidence of psychological withdrawal: subjects were shutting themselves off psychologically and were filtering or censoring what they said to the support personnel. The effect could be more or less pronounced, depending on the specific mission control team (it took four such teams to provide continuous support). Some teams had better styles for communicating. The researchers point out that nonoptimal communication styles can have adverse psychological effects as well as can restrict the flow of crucial information. One type of communication that did increase over time was crewmember requests for information about what was going on in the outside world.[46]

Spacefarers have had emotionally charged exchanges with mission control and on occasion have failed to follow instructions.[47] In one instance, begging off on the grounds that he had a head cold, an astronaut failed to don his required space helmet as the space capsule returned to Earth. In another instance, an astronaut removed biosensors intended to monitor heart function. Eugene Cernan reports that one early Apollo crew ignored instructions, broke rules, and regaled listeners with a stream of snide comments and complaints.[48] "For eleven days they did two things extraordinarily well," Cernan writes: "successfully performing every test and pissing off about everybody in the program, from grunt engineer to flight director." None of these astronauts ever flew again. But the most celebrated and controversial episode during the early years of the U.S. space program was the so-called Skylab rebellion.

In essence, NASA had planned every moment of every day for this crew's three-month mission. Headquarters' plans did not, however, take into account the time required to erect the apparatus or the difficulty of doing work in space. The crew considered it unrealistic to wake up at the designated time, go straight to work, break to eat, go back to work, have supper, and then return for sleep. At one point, the crew insisted on completing an experiment, rather than abandoning it in midstream just to start the next experiment on schedule, and they even took a couple of days off.

There are two interpretations of this. According to the more dramatic account, set forth in H. S. F. Cooper's book *A House in Space,* tensions mounted until the Skylab crew refused to complete assigned work and essentially staged a sit-down strike.[49] According to the milder account, the crew had worked very hard during its first few weeks in space, and at one point, fearful of cumulative fatigue, arranged to take a day off and adjust their work schedule.[50]

Hostility to and even occasional defiance of mission control may serve some useful purposes for a crew. The tensions generated within the group may be relieved by redirection to outsiders. That is, it might be better to become angry with someone hundreds of thousands of kilometers away than with the person strapped into the adjacent canvas couch. Furthermore, perceiving mission control as a common enemy can enhance cohesiveness within the group.

Jerry Linenger reports fairly severe conflicts with Russian mission control during his five months aboard Mir.[51] Although he was pleased with the harmony within the crew, he was disappointed with the manner in which the ground control interacted with the crew, breaking promises, yelling, and blaming the crew for everything that went wrong. He believed that ground control personnel couldn't be trusted and that, rather than providing the crew with support, they were the crew's nemesis. Meanwhile, as is common in cases of conflict, people on the ground saw the crew as the problem. As might be expected from prior research on conflict and cohesiveness, Linenger found that strained relations with support personnel strengthened the bonds among the crewmembers.

Refusing to wear a helmet, directing rude comments to the ground, and even participating in a "sit-down strike" seem rather far removed from, say, refusing to obey a direct order to return to Earth or pushing an unclad NASA official out of an air lock. In any case, there have been conflicts in space and there will be conflicts again. At some point, some of these conflicts may have grave consequences. We cannot hope to eliminate conflict, but selecting people with interpersonal sensitivity, as well as providing awareness training and skilled leadership that attends to both the work-related and emotional needs of the crew, should keep conflict from becoming too destructive.

CONCLUSION

Because space exploration is a group endeavor, researchers studying humans in space must take into account relationships within and between groups. Changes in crew size and composition give rise to new questions about personal and social adaptation to space. Larger crews allow spacefarers to do more with a greater margin of safety. Similarly, in comparison to crews whose members are nearly clones of one another, crews that contain a diverse group of people have a broader range of talent and offer more social stimulation.

On the other hand, increasing crew size and diversity raises issues of coordination. Different native languages may make it harder to communicate under conditions where communication is already difficult, and differing cultural preferences have implications for logistics and recreational activities. Moreover, people who are perfectly fit for space may find themselves handicapped by other crewmembers' prejudices.

The spaceflight environment affects essentially all basic social processes. Under conditions of microgravity it is more difficult to communicate nonverbally. Many of the conditions known to cause cohesiveness or heightened esprit de corps are associated with spaceflight. Because crewmembers are so highly dependent upon one another, we expect that social norms will be rigorously enforced. Whereas this may ensure conformity, it also has many other consequences, such as inhibiting novel ideas and leading to potentially devastating effects on members who fail to go along with the group. Isolation and confinement tend to heighten interpersonal tensions, both within the crew and between the crew and mission control. Once again we find that although space is a challenging environment, there are no showstoppers. Given the right attitudes and the right training, people have bright prospects for meeting the interpersonal stresses of spaceflight.

CHAPTER 9

AT WORK

In the earliest days of spaceflight, the basic requirement was to go into space and get back alive. Over the years, however, other tasks have been added. Now, spacefarers not only fly and maintain spacecraft, they launch and retrieve small satellites, conduct scientific research, construct space stations, and otherwise develop the infrastructure for the long-term habitation and industrialization of space. They also engage in less glamorous activities, such as repairing life support systems, vacuuming filters, packing garbage, preparing food, flattening out empty containers, and unpacking and stowing equipment. There is always more work to do than can be accomplished easily, and work assignments may be timed to the minute. Even when he was asleep on Mir, writes Jerry Linenger, he had electrodes pasted to his eyelids so that he could generate scientific data. "I wanted to complete all goals—no exceptions," writes Linenger, "and if I had to be something of a robot, so be it."[1]

On the space shuttle, pilot astronauts (briefly introduced in chapter 6) are responsible for making command decisions and for flying the spacecraft. These include the commander, who sits in the left front seat and who is in charge of the overall mission (something like the captain of a ship), and the pilot, who sits in the right front seat and whose role is analogous to that of the copilot in aviation. The commander is responsible for the safety of the vehicle and crew and for mission success. The pilot helps the commander control and operate the shuttle and may operate the remote manipulator, a robot arm, to deploy and retrieve satellites. Pilot astronauts, with their background as jet aviators, are the closest direct descendants of the "right stuff" astronauts of the 1960s.

Mission specialists provide technical support for the spacecraft and mission. They manage the habitat and its equipment, ensure that life support systems are functioning properly, handle logistical details, and maintain supplies. They work as flight engineers and are responsible for orbital activities: overseeing payloads and experiments, undertaking space walks, and, when the responsibility is not assigned to the pilot, controlling the remote manipulator. Although well versed in the overall mission requirements and goals, their specific responsibilities vary from mission to mission.

Payload specialists do the scientific or commercial work that justifies the mission. Payload specialists include earth scientists, life scientists, engineers, and others who have great expertise in a technical area crucial for achieving that mission's objectives. Most of today's payload specialists are research scientists, but a few have worked on commercial projects such as placing and retrieving satellites, making observations that benefit world agriculture or commerce, and trying out manufacturing processes. Future payload specialists may operate solar power stations, mine the lunar surface, produce commercial goods under conditions of microgravity, entertain tourists, and transport ore around our solar system. Performing under a variety of occupational titles, tomorrow's payload specialists could spend their entire adult lives working in space.

SPACEFLIGHT CONDITIONS AND HUMAN PERFORMANCE

Spaceflight conditions can make even the simplest tasks difficult. Inside the habitat, spacefarers have to work under cramped conditions. In transit, or in orbit, even staying put can be difficult thanks to microgravity. Space adaptation syndrome, cardiovascular deconditioning, and the other biomedical effects of spaceflight may undercut performance—after all, it's hard to maintain a high level of proficiency if one is sick or weak. To stay alive outside the habitat, spacefarers must wear bulky and cumbersome protective gear that makes it even more difficult to work.

Perception

Perception is based on a mixture of the "real world" stimuli that confront our senses, and the psychological processes that make these stimuli seem organized, coherent, and meaningful. The "real world" of the space environment differs in many ways from the one that we are accustomed to on Earth, and these differences affect our perception. I have already discussed how living under conditions of microgravity can make it difficult to gain a true sense of up and down. This can lead to brief periods of disorientation and to performance lapses that could prove dangerous under emergency conditions.

On the one hand, since the terrestrial atmosphere reflects light, even on a dark night people near settled areas gain some illumination from artificial light after it has bounced off of clouds. On the other hand, the terrestrial atmosphere absorbs a percentage of incoming light, with the result that in daylight high-altitude aviators and spacefarers work in outdoor settings that are approximately 25 percent more brightly illuminated than on our planet's surface.[2] Thus, outer space is characterized by sharper visual contrasts than we are used to on Earth. On Earth we are accustomed also to gradual transitions of light as the Sun rises and sets. On a satellite that orbits Earth on a ninety-minute cycle, the transitions are abrupt, something like emerging from a dark theater into broad daylight after a matinee. For spacewalkers, sharp contrasts and rapidly shifting shadows make certain kinds of work, such as erecting a truss, difficult. It may be less confusing to do this type of work during the orbital night, relying on weak illumination coming from flashlights.[3]

The absence of an atmosphere that absorbs and diffuses light suggests that vision should be very sharp. In fact, early reports suggested that people accomplished remarkable feats of acuity, such as identifying rivers and other terrestrial reference points, vehicles, and even groups of people at distances that would seem impossible on Earth. Yet, so far, these reports have not withstood scientific research.[4] Even though spacefarers may be operating in a very clear environment, several conditions work against good vision. These include changes in pressure within the eyeball brought about by increased fluid within the head, vibration of the spacecraft, and difficulties in accommodating rapid shifts in day-night cycles. Of course, if we move significantly farther away from our Sun or begin establishing bases on planets that have their own unique atmospheres, we will encounter new conditions that could have profound effects on our senses.

Circadian Rhythms

On Earth, our physiology and performance vary as a function of time of day.[5] These cyclical variations, evidenced in body temperature, heart rate, energy levels, and work output are known as circadian rhythms. Many of us attain peak energy and performance shortly after awakening and then gradually slow down as the day progresses. Because of these fluctuations, most of us are not at our best when we are roused from sleep in the middle of the night and expected to do something physically taxing or that requires an alert mind. Alertness and fatigue do not equally affect all tasks; as we tire, we lose the ability to do complex and demanding tasks much more rapidly than the ability to do simple or routine tasks, so it makes sense to face the most challenging assignments early in the day. These cycles are governed by internal physiological events and by external stimuli or cues, such as the position of the Sun, wristwatches and clocks, and social events such as mealtimes. These external reference points are known as *Zeitgebers,* which in German means "time givers."

People can run entirely on their internal clocks, but in the absence of external reference points they will gravitate toward 25.4-hour days. Usually, our internal clocks are "reset" by *Zeitgebers,* but if they are unavailable we drift out of synchronization with the rest of the world by several hours per week. Furthermore, because the 25.4-hour cycle is a statistical average (some people gravitate toward longer days and others gravitate toward shorter days) members of work groups can get out of synchronization with one another.

Normal sunrises and sunsets, perhaps the most important natural *Zeitgebers,* aren't available in underground bunkers, submarines, and spacecraft. Not only do orbiting spacefarers experience sunrises and sunsets at ninety-minute intervals, there are many other factors that can disrupt their internal clocks. Russian spacecraft are not in constant communication with the ground. Consequently, cosmonauts may be aroused at those times when two-way communication is possible. The times at which Mir passes over different communications posts differs from day to day, making it even more difficult to remain synchronized.

As we move toward either the North or the South Pole the lengths of the days and nights throughout the year become increasingly variable, culminating in one six-month "day" and one six-month "night" at each of the poles themselves. The lengths of days and nights vary from planet to planet. One comforting aspect of Mars is that its period of rotation is almost identical to Earth's, so we can expect good synchronization between internal and external time.

There are several ways of offsetting desynchronization. We can set the clocks onboard so that the space crew operates in the same time zone they operated in at their point of departure from Earth. Thus, U.S. spacefarers operate on Houston or Cape Canaveral time, while Russian spacefarers operate on Moscow time. For a few days before liftoff, spacefarers may acclimate to the host country's time zone, so, for example, astronauts on Mir will be synchronized with Moscow rather than Houston time. Second, shutters that block out light or darkness and lights turned on or off can mimic sunrises and sunsets. Third, as Jack Stuster points out, we can discourage "free cycling" by encouraging people to go to bed and arise at their normal times.[6] He adds that we should also urge people to eat together and engage in other social activities that provide useful reference points. Through such simple expedients as forcing people to fall in at 0700, eat breakfast at 0800, and avoid naps, leaders of early polar expeditions were able to combat the problems associated with a seemingly endless night.

Working in Microgravity

Because we are so used to normal gravity, we don't pay much attention to its effects on our performance, except, perhaps, when we are puffing uphill. Gravity gives us traction to move from one place to another, and, once we are there, it helps keep us firmly in place. In microgravity we lose this advantage. Working under conditions of microgravity is similar to working on an extremely slick floor. Spacefarers are forced to use handrails, tethers, and other devices that compensate for poor traction. Skylab astronauts' footwear had triangular metal cleats that could be inserted into triangular sockets on Skylab's floor. This offered firm footing but complicated the process of moving around the workplace. Cosmonauts have slung ropes or cables through space stations to pull themselves from one place to another, much as one might pull oneself along a rope in a swimming pool.

As many people have learned through experience, it is difficult enough to keep track of small tools and parts on the two-dimensional surface of a workbench, never mind under conditions where items can simply drift out of sight. Everything from pens to very expensive cameras have been lost this way. Many early crews left everything out until time to return to Earth and then frantically stowed everything at once. Stowage design improved dramatically for Skylab. Today's spacecraft include a wealth of specialized storage compartments, and the old adage "A place for everything and everything in its place" takes on new significance. Under conditions of microgravity, belts, tethers, pockets, snaps, and acres of Velcro are the order of the day.

Space Suits and Extravehicular Activities

Extravehicular activities (EVAs) are strenuous and dangerous. Most run four hours, and some have gone for eight hours. In addition to the time spent on the EVA itself, there is time for preparation and wrap-up, both of which are done inside the air lock. Outside the habitat, moving even short distances is difficult when encumbered by a bulky space suit. Relatively early in the development of spacecraft, handholds were added to the exterior to make it easier for spacewalkers to scamper about. Later came small, personal propulsion units using tiny rocket engines or compressed gas.

Space suit design is a fine art that requires resolving many dilemmas.[7] Inspired by suits of armor and deep-sea diving attire, the space suit must be strong enough to protect the occupant, but flexible enough so that he or she can move around. An underinflated, soft space suit is easy to move around in, but if too underinflated the suit will not keep the person alive. Overinflating a space suit is in some ways safer but makes it difficult for the occupant to move his or her arms or legs. Having only a small faceplate makes it difficult to see what one is doing, but since holes of any type are points of structural weakness, having a large faceplate increases risk. The greater the number of joints and movable parts the easier it is to move around, but the more opportunities there are for the space suit to develop leaks.

Despite preconceptions of space as a frigid void, dissipating heat is the greater problem. The outside temperature may be well over a hundred degrees. Inside the space suit, the hard-working spacefarer is giving off more heat. Not only is this uncomfortable and smelly but dripping sweat and a foggy visor make it difficult to see, a situation that cannot be remedied with a simple swipe of the hand. To help offset this, spacefarers wear undergarments containing tiny tubes that have cold water circulating within them.

Working within a space suit requires more time and energy (oxygen, water, and food) and leads to much greater fatigue than working in regular clothing. Operating equipment or conducting work that involves intensive use of the hands and fingers fatigues even fit astronauts in relatively short order. Early EVAs were tremendously exhausting and caused spacefarers to sweat quarts of water, which they poured out of their space suit boots after they squeezed themselves back inside the spacecraft.

Designers seek lightweight, flexible suits for future missions. Particularly desirable are increased flexibility for wrists, hands, and fingers and feedback devices that give at least the illusion of touch. In the future, hard space suits modeled after suits of armor will allow occupants to work at normal rather than reduced atmospheric pressure, without the "ballooning" characteristic of the soft space suits.

Shuttle space walks are related to specific tasks that are set in advance. They can be carefully planned and well rehearsed. On space stations, space walks often have to be improvised as it becomes necessary to make some sort of emergency repairs. The spacefarer may not know what to expect until he or she exits the hatch. Furthermore, moving from place to place can be extra dangerous on a space station because the various antennas, solar panels, and experiment packages have sharp corners and can obscure the view. It is not that difficult to become disoriented, and more than one spacewalker has wondered about finding the air lock to get back in.

Role Loading

During the Spanish-American War, Frederick W. Taylor carefully analyzed tasks in terms of their component parts.[8] By eliminating unnecessary actions and streamlining those that remained, he and his followers achieved prodigious gains in productivity. For example, by doing away with the need to bend over to pick up bricks (by storing them on adjustable scaffolding) and reducing from eighteen to five the number of motions needed to lay a brick, the average bricklayer's output was tripled.

At a launch cost of thousands of dollars per kilogram it seems very reasonable to assign astronauts heavy workloads. It also makes sense to apply principles of motion economy (regarding the arrangement of the workplace and the design of tools and equipment) to achieve high efficiency. And indeed, there have been many time-motion studies in anticipation of the ISS. However, in our zeal for results we should not overlook another of Taylor's findings, the need to minimize cumulative fatigue. He found that workers who were given rest periods outperformed those who continued at full speed throughout the day. The most productive workers were those who were free of load at least 50 percent of the time.

Thus, although we welcome attempts to increase productivity through simplifying tasks, we should not support attempts to increase productivity through endless additions to already long lists of activities. Such lists can be counterproductive and, as noted in chapter 8, they adversely affected the final Skylab crew. The scheduling was so tight that not only was there no free time, the astronauts were expected to start new projects on schedule even if it meant abandoning other projects in midstream. This led to friction between the Skylab crew and mission control.

When programming spacefarers' time it is important to remember that (1) tasks easily accomplished under normal conditions may be very difficult to accomplish in microgravity or while wearing a space suit, (2) the same performance levels that can be maintained on brief missions are difficult or impossible to sustain on missions lasting weeks or more, and (3) spacefarers must have enough time to pursue unexpected opportunities. Schedules should leave time to follow exciting new leads.

THE SPACEFARER'S TOOL KIT

Although the view that people are the only tool users was demolished long ago, we humans seem unparalleled in our ability to create artifacts that permit us to survive, flourish, and amuse ourselves. Tiny implements and heavy earthmoving and construction equipment represent a continuation of a long-term trend. Pulleys, levers, and wedges arranged in endless variety supplement the strength of our arms and increase the mobility conferred by our legs. In the past, water and wind augmented our muscles; today our strength is further amplified by the Sun and by fossil and nuclear fuels. In addition to extending our physical powers, we have begun, on a serious basis, to enhance our intellectual powers with machines that assemble, store, and process information. These range from such simple devices as paper and pencil to giant supercomputers capable of storing enormous amounts of information and processing it at lightning speed.

Space has always been a high-tech environment, but it will be increasingly so in the future. Consequently, we must ask ourselves questions about people, tools, and machines. These range from fairly simple questions about the design of workplaces and tools to the use of intelligent machines that are programmed to make decisions and then take action if necessary (for example, by shutting off an overheating motor).[9]

Work Spaces

Work spaces must be designed to accommodate two realities of spaceflight: cramped conditions and microgravity. To address the first of these issues, human factors engineers identify "activity envelopes"; that is, the amount of space that people require to complete designated tasks (see chapter 5). If a person has to bend over, twist, and stretch to complete a job, then designers must allow enough space for these activities to occur. Plans must call for accommodating people of all sizes, ranging from all but the smallest (5 percent) women to all but the largest (5 percent) men. Additionally, the design of work spaces must take into account changes of size and posture under conditions of microgravity. Because it is no longer compressed by gravity, the human body tends to elongate or stretch perhaps five to seven centimeters. This and the microgravity-induced "space slouch" reminiscent of a fetus floating in amniotic fluid affect both reach and the ability to position the arm and hand accurately. For this reason, orbiting workstations have to be designed differently from workstations on Earth.

Spacecraft that have many successive crews should allow for a lot of people to operate onboard systems. Flight data files, notes John Schuessler, contain handbooks for operating and troubleshooting every system. Some of the flight data files are paper, some are card stock, and some are electronic. Nevertheless, it is necessary for all of the crews, as they rotate through the habitat, to be able to operate any system quickly and properly. One of the techniques that was effective on Skylab was to etch a schematic drawing right on the face of the control panels that graphically showed connections between the motors, relays, valves, and other components connected to those controls. Going to the flight data files was necessary only under extreme conditions.[10]

Storage is another issue. Russian space stations are cluttered affairs with every nook, cranny, and module stuffed with equipment and supplies. An advantage of the shuttle is that it has to carry only the equipment needed for one specific mission. Furthermore, this equipment can be "combat loaded" so that it can be retrieved in the order in which it is to be put to use. The crew is highly trained in its use. On the ISS, as on Mir, crews will have to use old equipment repeatedly and perhaps in new ways, and will have to learn how to operate new gadgets that arrive during the mission. Successful operation may require many radio conversations, with the spacefarers questioning inventors to learn how something works. As a result, we would not expect work on the ISS to match the fast pace on the shuttle, and despite fastidiousness we would expect the ISS to develop a cluttered look.

Basic Tools

Many of the tools found in your garage tool kit would work fine in space. The shuttle tool kit contains common household tools such as hammers, screwdrivers, wrenches, pliers, wire cutters, and duct tape, all very useful for on-the-spot repairs. Occasionally, spacefarers have fabricated implements to save the day. The astronaut Mike Mullane reports how crewmembers once cut a piece of plastic in the shape of a flyswatter that enabled them to flip a switch to restart a dormant satellite. Unfortunately, the switch was not the problem![11]

Specialized tools for use in space must be lightweight, reliable, and compact. They must be easy to use within the tight confines of the habitat despite microgravity and while wearing bulky protective gear. Tools developed for spacefarers include a zero reaction power tool that looks like an ordinary hand drill but does not have a twisting reaction force (thus making sure that it is the screw, not the spacefarer, that rotates), a ratchet screwdriver that is easy to use even while wearing heavy gloves, a spring-driven hammer that works well under conditions of microgravity, and an aluminum fingernail that fits over the finger of a space suit glove and is used for picking up small objects.[12]

Special care must be given to the development of controls such as dials, gauges, and switches. Gauges should be easy to identify and have highly legible displays; switches should be large and easy to use without inadvertently tripping their neighbors; and rotary dials should be easy to grasp and rotate and not require fine alignment for proper operation. Even interior equipment must be designed so that it can be operated from inside a space suit, as could become necessary following depressurization. Additionally, controls should be standardized, so that the location and operation of controls doesn't change as the spacefarer moves from one place to another. (Because so many nations are involved in building the ISS, standardization is very difficult.) The idea is to provide the worker a little more latitude, an extra margin of safety, in recognition of the difficulty of working in space.

Astronauts have tended to work by manuals and checklists, while Russian cosmonauts tend to work by knowledge in their heads. This too reflects some of the differences between short-term and long-term spaceflight. Because shuttle trips are brief, it is possible to set out and follow a very specific plan. Space station missions require more improvisation. Thus, in comparison to shuttle astronauts, space station cosmonauts have been forced to figure things out as they go along.[13] ISS residents will have wrist-mounted checklists with liquid crystal displays, but will still be required to engage in last-minute problem solving and improvisation.

PARTNERING WITH INTELLIGENT MACHINES

Work, notes Gregory Stock, involves a three-part sequence. These are sensing, analyzing, and responding.[14] Each of these activities can be accomplished by people or machines. Video cameras, microphones, computers, levers, and wheels are the mechanical equivalents of eyes, ears, brains, arms, and legs. As technology becomes increasingly sophisticated, and machines become better at sensing, analyzing, and responding, there are more and more opportunities to take the human out of the loop. This raises the issue of how we should divide work between people and machines, and the kinds of problems that we are likely to encounter when we try to do so.

Assigning Tasks to People and Machines

Mary Connors suggests that the best way to approach human-machine partnering is to ask what machines do best and what humans do best and assign work accordingly.[15] Automated systems, she notes, reliably handle data that are properly defined and predictable. They can and often do operate with great speed and perform repetitive tasks as often as needed. They have tremendous stamina and are not disrupted by environmental conditions that make people fearful or angry. However, the data must be precisely defined. E-mail is delivered with great speed and accuracy providing that the address is *exactly* right. Unlike your neighborhood postal carrier, the computer won't compensate for a minor misspelling of your name or the transposition of two numbers in your street address.

Humans, Connors adds, have compensating strengths and weaknesses. People are able to deal with ambiguous, "fuzzy," or missing data and can organize information in many different ways. Oftentimes, they respond very effectively to unanticipated and changing conditions. Unlike machines, humans can think "outside the box." People are "best at creative work, they deal well with the unexpected, they are able to take advantage of serendipitous events, to rely on intuition, and to demonstrate common sense."[16] On the other hand, when humans perform repetitive tasks, errors and mistakes tend to creep in. Humans are subject to boredom and fatigue, and emotions such as fear and anger affect their performance.

In his discussion of interstellar travel, Raymond J. Halyard envisions a small human crew augmented by specialized robots, androids, and a colossal main computer.[17] Much of the work in space has been and will be done by robots that serve as stand-ins or partial stand-ins for humans. Robots that have landed on the Moon, Mars, and Venus have collected and analyzed samples and, through telemetry, returned photographic and other data to Earth. Some of these devices move around and carry sophisticated automated laboratories that conduct biochemical assays.

Any task that can be assigned to a robot—such as the analysis of soil, rock, and ice samples—leaves the spacefarer free for other duties. Robots multiply human strength. Spacefarers may lack the strength to move large objects in and out of the shuttle's cargo bay, but they can easily work the controls of a large and powerful remote manipulator. Robots can be programmed to precisely locate a payload, module, or system with a degree of speed and accuracy that would be difficult for a human. Robots allow us to do work under dangerous conditions, such as handling dangerous chemicals or operating in lethal environments. Using robots in poisonous atmospheres or under "hot" radioactive conditions are examples of this. Robots, not humans, move the control rods that keep nuclear reactors functional. Because robot explorers do not need life support systems, they are relatively cheap, and we are willing to take far greater risks with robots than we are with people.

Ronald D. Jones foresees "bootstrap" robots that will be valuable for interplanetary missions.[18] They could be designed to "exploit indigenous materials, extracting from the surface oxygen, metals, glass, water and other items needed by the base." These robots could be sent in advance of humans, perhaps to stockpile oxygen and water or manufacture fuel for the return trip. In the distant future, we might envision workshops utilizing nanotechnologies able to recycle broken equipment and, from the molecular level on up, fabricate replacements.[19]

Halyard's androids would look like humans and would approximate human intelligence. Although they would be blunted emotionally, they would be pleasant enough and have human mannerisms. Capped intelligence and blunt emotions would prevent a "revolt of the androids" a few generations out on a multigeneration mission.

The colossal megacomputer, used to store incalculable amounts of information and solve very complicated problems, would be infinitely more intelligent than the human crew. The relationship between the computer and the human would be analogous to the relationship between a human and a slug. To avoid threats to humans, Halyard suggests, the supercomputer would have no contact whatsoever with the androids or the specialized robots.

Michael C. Greenisen and Victor Reggie Edgerton observe that because of the rapid growth of intelligent systems in space, the spacefarers' primary job is one of systems management and decision making.[20] As more and more of the actual work is accomplished by automated systems and robots, we rely less on humans to perform physical work and more and more on them to monitor systems, make decisions, and then adjust the system accordingly. Furthermore, people's abilities to diagnose and repair space systems means that the systems themselves do not have to be quite so complex, redundant, and expensive as they would if they had to be entirely self-maintaining.

Trust

A perfectly reliable device may fall into disuse simply because people question its ability to operate properly. Reluctance stems from uneven track records for automated systems, convictions that the system can't match the performance of a human operator, and those forms of inertia (rooted in social conventions or in personal habit) that keep us doing things the way we always did in the past. Furthermore, we may reject a system because we believe that it cannot possibly be working right, even when it is. Consider part of William Manchester's account of the hurricane that devastated the eastern coast of the United States in 1938.[21] On September 21, a Long Islander received a barometer that he had ordered from a store in New York City some days before. After it arrived in the morning mail, the new owner discovered that the needle rested below 29, where the dial read "Hurricanes and Tornadoes." He shook it and banged it but the needle did not move. He repackaged it and then drove it to the post office to send back to the store. When he returned home he discovered that his house had blown away.

On the one hand, we should not allow ourselves to be lulled into insensitivity. Machines, even those that appear to be working well, can and do fail. On the other hand, we should not be so suspicious that we either refuse to let the machine do its job, or needlessly duplicate its efforts. Useful here are "reality checks," procedures that make it possible for the operators to verify that their machines are functioning properly.

Who's in Charge Here?

Many if not all spacefarers are "take charge" types of individuals used to being in control of situations. One indication of this is the early astronauts' reluctance to accept proposed versions of the Mercury craft that could not be controlled by the occupant. Instead they lobbied successfully for viewports and an imitation joystick in the form of a side-arm controller that provided at least a little control over the flight. We know from many a study in psychology that there are advantages to feeling in control of a situation. People who feel that they are acting, rather than acted upon, tend to have higher self-esteem and attain higher levels of success, because they energetically and proactively bring about the conditions that they desire. People who believe that they are at the whims of fate develop a sense of helplessness and don't try very hard. In addition to energizing performance, a sense of control helps people tolerate difficult situations.

CONCLUSION

Space is a demanding work environment where the design of tasks and tools must reckon with weightlessness, vacuum, and other environmental conditions. Work that is relatively simple under everyday conditions is challenging in space. The need for orientation and anchoring in microgravity, the requirement to work in cramped spaces, and the special challenges of extravehicular activity must be met. Work will be very different as we make the transition from the shuttle to the ISS and beyond.

Shuttle missions are brief and have carefully defined objectives. This means that shuttle astronauts can be highly trained for their tasks. They can rehearse again and again. Because the time in orbit isn't all that long, they can be armed with detailed instructions and can maintain a rapid work pace. Moreover, the shuttle carries everything it needs to meet a specific flight's objectives. There is no need to store equipment that has been used earlier or won't be used until later. This means equipment is relatively accessible. All space missions involve housekeeping duties, but because shuttles are away from Earth for brief periods these housekeeping duties are relatively light. On the shuttle, relatively little time is spent in maintaining and repairing equipment. If something important breaks down, the shuttle can land to have it fixed.

Space station missions are lengthy and have multiple objectives. Because duties of space station residents are broad, it is less possible to provide these people with the same depth of training given shuttle astronauts. People on space stations cannot maintain the fast pace of those who make only brief forays into space, and they require greater balance in their lives.

Some space station assignments may be received after the spacefarers have left Earth. For this reason they cannot rely so heavily on manuals and checklists. The space station must store all of the materials needed for months at a time. It may be difficult to pack and store equipment efficiently, and the habitat is likely to develop a cluttered look.

Because the space station is larger than the shuttle and its service life may be measured in years, it will require more housekeeping and maintenance work. Unrehearsed space walks may be required to fix an air lock or to maintain equipment attached to the outside of the craft. There is no landing a space station so that a ground crew can make repairs. All repairs must be made in orbit, using the equipment at hand. The Mir experience will become the ISS experience.

As more and more people move into space, we must continue to give them the best tools we can, ranging from aluminum fingernails for retrieving small items to huge robot cranes and automated manufacturing systems. And, as space manufacturing and tourism appear, we will have to provide tools for an ever-expanding range of activities. With the proper design of restraints and tools, we should be able to do anything in space that we can do on Earth, and then some.

CHAPTER 10

MISHAPS

Given the sheer size and complexity of space missions coupled with the lethality of outer space, our safety record during the first half century of exploration is remarkable. Nonetheless, amid a long list of triumphs are occasional tragedies that involve the loss of Russian and American life.

On March 23, 1961, the cosmonaut trainee Valentin Bondarenko died after he unwittingly tossed a small piece of alcohol-soaked cotton, used for wiping off electrode paste, onto a hot plate. A fire erupted in the oxygen-rich environment of the simulator, and rescuers could not open the door quickly because the high atmospheric pressure in the cabin kept it from swinging inward.[1]

On January, 27, 1967, during the Apollo 1 simulation mission, because of a door that was secured by a ring of bolts and took seven minutes to open, rescuers could not save three astronauts—Roger Chaffee, Gus Grissom, and Edward H. White—after a fire that broke out due to faulty wiring. This event, which closely resembled the disaster that had befallen Bondarenko, might have been avoided if the Soviet Union and the United States had openly shared information at that time.[2]

On April 24, 1967, a parachute failed to deploy during the landing of the Soyuz 1 spacecraft, causing the craft to crash into the ground at over three hundred kilometers per hour, burst into flame, and kill the occupant, Vladimir Komarov. According to a Russian engineer, Victor Yesikov, Soyuz 1 had been launched prematurely and had had many problems, leading Komarov to radio angrily, "Devil machine, nothing I lay my hands on works."[3]

On June 30, 1971, a faulty valve arrangement allowed air to escape from Soyuz 11, asphyxiating the cosmonauts Georgi T. Dobrovolsky, Victor Pataseyev, and Vladislav N. Volkov as they descended to Earth. This had been the first crew sent to Salyut 1, the world's first space station, where they had lived and worked for about three weeks.[4]

On January 28, 1986, about a minute after the space shuttle *Challenger* was launched, O-rings used as seals in the solid rocket boosters burned through, allowing escaping gasses to ignite the fuel in the external liquid-fuel tank. This occurred within the view of spectators at Kennedy Space Center and was clearly captured by telephoto lenses. The resulting deaths of Dick Scobee, Michael J. Smith, Judith A. Resnik, Ellison S. Onizuka, Ronald E. McNair, Gregory B. Jarvis, and Sharon Christa McAuliffe shocked the nation and led to a temporary suspension of the shuttle flights.[5]

The Russians, who land their cosmonauts by parachute, have had some notably rough landings in addition to the one that killed Komarov.[6] Cosmonauts have endured forces of up to 10-Gs when they were forced to land at very sharp angles. They have been temporarily lost in mountains, have been dunked in freezing lakes, and have landed ominously close to the Chinese border. In these cases, survival training made a difference and the stories have happy endings.

Perhaps the best known close call in the U.S. space program came during the Apollo 13 mission. The explosion of an oxygen tank on the outward leg of this journey to the Moon led to a loss of power that put propulsion and life support in jeopardy. Through extraordinary efforts of ground-support crew as well as the in-flight astronauts, Jim Lovell, John Swigert, and Fred Haise, the crew used the lunar landing module as a safe haven, completed a swing around the Moon, and repowered the craft for a safe splashdown.[7]

As we review a partial litany of things that can go wrong we must remember that people can and do recover from structural or mechanical failures and from their own mistakes. Indeed, there may be hundreds of equipment failures or human mistakes for each failure or mistake (or, more typically, combination of failures and mistakes) that leads to the loss of the spacecraft or crew.

FAILURES AND ERRORS

The sociologist Charles Perrow has identified six factors that contribute to accidents.[8] He summarizes these with the acronym DEPOSE:

design, equipment, procedures, operators, supplies (and materials), and environment. For example, an accident could come about when an improperly designed rocket shakes apart (design), an oxygen regeneration system fails (equipment), an important step is inadvertently left off a checklist (procedures), a crewmember inadvertently hits the wrong button (operator), the craft runs out of fuel prior to reentry (supplies), or the crew receives a fatal dose of solar radiation (environment).

Analyses of accidents tend to focus on the operator. Even in the case of Apollo 1, someone had the temerity to suggest that crewmember Gus Grissom (who had been reclining quietly in his loungelike chair) had somehow managed to kick the wires (that were located behind him), thereby causing the sparks that led to the conflagration. In aviation, operator error causes only about half of the accidents. Most of these do not involve problems of control ("stick and rudder" accidents) but rather faulty decision making and poor communication among crewmembers.[9]

Several biases contribute to the tendency to blame the operator. The manufacturer who builds the device is eager to avoid corporate responsibility. Accepting blame could cause a whole program to come to a halt, perhaps by grounding all models of a particular aircraft or forcing expensive retrofits. Similarly, denying that bad engineering or sloppy workmanship were at fault may reduce legal judgments in favor of grieving relatives. Moreover, our legal system tends to emphasize "proximate cause," or the most recent event just prior to an accident. In a chain collision, for example, when someone else hits your automobile and sends your car careening like a billiard ball into yet another person's car, you may find yourself liable. In theory, you were the last person in the chain of events who could have prevented damage to the next person's vehicle. Likewise, if a sixteen-hour shift forces an operator to fall asleep on duty, overseers may neglect the outrageous length of the shift and focus on the operator's performance lapse. C. O. Miller points out that accident investigators sometimes conduct shallow investigations that stop after scrutinizing the operator's behavior; by failing to look for other contributing factors, they strengthen the impression that the accident was due to operator error.[10]

Miller adds that most accidents reflect a chain of events or have multiple causes. When we are tempted to pin the blame on the operator, we should remember Elwyn Edwards's admonition that most human errors are not "due to sudden illness, to suicidal tendencies, to willful neglect, or to lack of basic abilities." Instead, they arise from "temporary breakdowns in human performance" that would have been inconsequential if designers had paid more attention to human characteristics and skills.[11]

Psychological Factors

Imperfect training, excessive workload, fatigue, lapses of attention, and stress are among the many causes of operator error. In aviation, spaceflight, and other high-tech environments, errors fall into three categories.[12] Display-related errors occur when a pilot fails to monitor instruments, does not notice a warning light or other indicator of systems failure, misreads an instrument, becomes disoriented, or succumbs to an illusion. Control-related errors involve inappropriate action. Such "stick and rudder" mistakes include throwing the wrong switch, failing to adjust a dial with sufficient accuracy, pushing forward on a control when one should pull backward, or inadvertently hitting a button with one's elbow.

Computers have added set-up and data-entry errors to the list. For the past few decades we have had to worry about operators misprogramming a computer or entering the wrong numbers. Such errors can be very difficult to detect and correct because it can take a long time for them to become evident. However, there are ways to incorporate safeguards. One example is to instruct aircraft flight computers to reject destination coordinates if there is insufficient fuel to get there.

The number of displays and controls increased steadily as we moved through two world wars, entered the age of jet planes, and then entered the age of spaceflight. Today's aviators and spacefarers are surrounded by warning lights, gauges, video displays, switches, levers, and dials, many of which are hard to see or difficult to reach. The shuttle has over two thousand displays and controls. The end result is that there can be a heavy demand on the operator's attention and ample opportunity for mistakes. Various solutions—for example, adding warning lights or tones—can make the problem worse if they add to the perceptual clutter. Offering summary information, a video display may make it easier for the operator to grasp the current state of the system, but when we aggregate information we lose potentially important detail. The question of how much information to provide and how to present it perplexed human factors engineers during the last fifty years or so and will continue to do so as we move farther into space.

Both fatigue and workload are important considerations. Even exhausted people may be able to do simple things, like pounding nails. What they lose is the ability to integrate information from several sources at once and perform tasks with several components. Tired people play close attention to some parts of their jobs but not to others, as when a sleepy driver keeps a close eye on the road but fails to check the instruments on the dashboard. As we get tired we can still perform certain subroutines, but we find it increasingly difficult to execute the overall program.[13]

Even as having too much physical work to do may slow a spacefarer's progress while assembling a space station, having too much mental work to do may undercut a person's ability to scan the environment, integrate information, and respond appropriately under emergency conditions. While it may be intuitively obvious that overworked people are prone to make mistakes, underworked people also are prone to errors. This is because a bored person may not pay careful attention to the task at hand, and do sloppy work.[14] Under conditions of extreme and prolonged boredom, people may experience odd sensations, feel detached from reality, even hallucinate. Pilots on very long overseas flights have reported a sense of detachment from their bodies, including a vivid sense of being outside the aircraft looking in, watching themselves fly the plane. Then there is the "giant hand phenomenon," which occurs when a pilot feels as if the controls are not responding properly to his or her touch. It is as if a "giant hand" is pushing the stick up and to the left.[15] Fortunately, pilots who reposition their hand on the stick regain a sense of control.

Sometimes when people appear to be doing their jobs, they are only going through the motions. In some of these cases, notes psychologist Ellen Langer, people function on the basis of ingrained habits and are oblivious to warning signs that the situation is changing.[16] Such automatic behaviors are a form of mindlessness, and can have catastrophic consequences. Langer points out that, in one case as a pilot and copilot mindlessly went through a checklist, they noted that the plane had been properly deiced (Check!) when it had not been. The result was that the overweighted plane crashed into a bridge immediately after taking off from Washington's National Airport. Mindful people attend to what they are doing, watch for critical changes in the environment, and notice problems before they get out of hand.

Dangerous situations can occur very suddenly ("They never knew what hit them!") or evolve slowly over time. Emergency situations, especially those that are unexpected and sudden, are likely to be associated with heavy mental workloads.[17] Spacefarers may be over-loaded by warning indicators, including flashing lights, buzzes, hums, and whistles. They may experience tremendous time pressures, resulting in part from the current state of the emergency and in part from recognition that if they don't do something right away the situation will deteriorate. Furthermore, the costs of extricating the crew from the emergency (such as by aborting the mission) may be very high. Unfortunately, when people are forced to act in haste, they may do something foolish or choose the right response but execute it imperfectly.

People's propensity to take risks is another consideration. On the one hand, if nobody had been willing to take chances, humans would never have left Earth. The early astronauts, for example, enjoyed competitive activities, heavy partying, rough play, and dashing around in Corvettes. On the other hand, unnecessary risks can increase danger. Sidney Blair points out that we may be better off choosing relatively conventional people for routine service in isolated and confined environments.[18] He reasons that space stations, polar bases, and underwater habitats are dangerous enough without the antics of people who take risks for their own sake.

The Russian compensation system, which links pay to performance, could encourage excessive risk taking. Under this system cosmonauts are fined for failing to fulfill their contracts, or awarded bonuses for extra work such as a space walk.[19] The bonuses may be substantial and quite valuable to the poorly paid cosmonauts. Linking pay and performance is one of the most powerful tools in the manager's arsenal, and it works fine when employees are doing piecework on an assembly line or selling vacuum cleaners door-to-door. This compensation system may be less appropriate if the prospect of monetary gain prompts people to take otherwise unnecessary chances.

Small-Group Factors

Some errors flow from the ways that different crewmembers interact with one another. As H. Clayton Foushee and Robert L. Helmreich point out, each crewmember can have excellent technical qualifications, but a lack of coordination among them sets the stage for performance lapses and accidents.[20] Leaders may fail to define responsibilities, to delegate tasks, or to assign priorities. They may neglect to monitor subordinates' performances or give prompt, accurate feedback. Each crewmember may meet their individual assignments, while no one is attending to the big picture. Crewmembers may express themselves in less than successful ways, with the result that other crewmembers do not take appropriate action.

Foushee and Helmreich add that some accidents have come about because crewmembers withheld crucial information from one another. Their reticence stems from a lack of trust in the system that gave them the information, concerns that the people with whom they want to communicate are dealing with even more important issues, or fear of ridicule or censure if it turns out they are wrong.

Other problems occur when crewmembers get together in a group to reach decisions. Highly opinionated people or those of high rank may dominate a discussion, even though they do not necessarily have the best ideas. Sometimes groups slip into *groupthink*, a mode of group decision-making, whereby efforts to maintain morale and amiability work against careful analysis and critical thought.[21] Through expressing great confidence in the leader and the group, protecting the leader from criticism, openly agreeing with one another (despite unspoken reservations or doubts), and ignoring or misinterpreting information that runs counter to prevailing opinion, the group reaches a decision that tells us more about how they wish things would be than about how things actually are. Under these conditions the group is likely to be so sure that it is right that it fails to establish contingency plans. Groupthink is particularly likely when the group is highly cohesive, when it is dealing with a crisis, and when it cannot (or will not) discuss the problem with outsiders. This kind of thinking probably contributed to the fateful decision to launch *Challenger* despite warning that the cold temperatures could cause serious problems with the O-rings.[22]

Crews should keep in mind that newcomers or outsiders may bring special strengths to problem-solving sessions. Newcomers, who are anxious to prove their skills, may be more motivated than experts who have dealt with similar problems in the past. Experts tend to rapidly assess situations and then move forward with preprogrammed solutions that are not necessarily the best. The neophytes' fresh perspective allows them to see things that experts do not, and they may be better able to think "outside the box" and find new solutions.

Cockpit resource management (CRM) refers to utilizing aviation and space crew capabilities to the fullest.[23] Effective commanders recognize that crewmembers may have important information, and they encourage them to communicate this information in clear and effective ways. Good management requires an authority structure that promotes communication and coordination, standardized procedures, training and evaluation programs that have consistent performance standards, and schedules and procedures that do not themselves add to the stress.[24] CRM programs try to improve performance on the flight deck by developing interpersonal relations, leadership styles, and communication patterns to improve communication, situation awareness, and decision making.[25] CRM has gone through several phases, and the most recent version is oriented toward safety.[26] CRM tools are developed within four aerospace cultures: national culture (for example, the United States), professional culture (astronauts), organizational culture (NASA), and safety culture. CRM is not a panacea, and certain principles, such as open communication, may be difficult to implement in certain cultures. Crewmembers from some cultures expect to keep their distance or to follow their commander's orders without question.

Organizational Factors

Sometimes the seeds of disaster are found in the imperfections in organizations that plan and manage space missions. Like operator error, managerial error does not necessarily reflect incompetence or willful neglect. It can stem from the organization's design, history, and culture, as well as individual and collective performance lapses.

The story behind the *Challenger* explosion is complex (after all, it takes a two-story warehouse just to hold *Challenger*-related documents) and there are many plausible interpretations. As Diane Vaughan explains it in *The Challenger Launch Decision*, the O-rings that were supposed to seal the various segments of the solid rocket boosters (SRBs) had been considered a potential weakness, and the contractor, Morton Thiokol, was particularly concerned that they might fail given the cold weather in Florida at the time scheduled for liftoff.[27] The low temperature, the company believed, could keep the rings from expanding fast enough following ignition, thus permitting hot gasses to escape from the side of the SRBs. Some Thiokol engineers registered their concern with NASA officials, but this protest did not cause higher NASA management to stop the countdown.

Immediately following launch, some O-rings completely failed but the burned residue kept the joint sealed temporarily. As *Challenger* rose from the pad it was subjected to forces that twisted and bent metal, only slightly, but enough for the residue to disintegrate. This allowed the flame to escape from the joint and ignite the liquid fuel in the huge external fuel tank. Given that some engineers foresaw great danger and warned NASA officials, how could this happen?

Vaughan points out that although it is easy to see managerial misbehavior we must try to look at the situation as NASA managers interpreted it at the time. They were attending to many worries, not just renewed complaints about the O-rings. NASA engineers thought they understood the O-rings: the seals could hold despite burning, but in any event there seemed to be enough redundancy built into the system that if one or two failed, other O-rings would hold. Nobody broke any rules. The sense that the risk wasn't all that great coupled with forceful personalities, red tape, and bureaucratic foul-ups led to final approval of the launch.

NASA is a classic bureaucracy with a clear-cut hierarchy, highly specialized workers, a welter of rules and regulations, and a penchant for written records. People who have important information to share may not be able to make contact with the people they need to see, and alarming or discouraging information may have trouble making it up the ladder to top management. Critical information may go to the wrong desk or be misinterpreted by superb administrators who have become somewhat rusty as engineers. The same bureaucratic characteristics that make an organization function effectively under routine conditions impair that organization's ability to respond quickly and effectively in emergencies.

Sometimes production pressures encourage risky courses of action. A felt need to "get on with it" means that some operations are commenced prematurely, with fervent hopes and prayers standing in for careful preparation. The Russian engineer Victor Yesikov, who later emigrated to the United States, observes that some early Russian launches were timed to serve propaganda purposes; for example, to coincide with an international meeting or to commemorate an important historical event. Vladimir Komarov was sent aloft in Soyuz 1 even though it had failed several tests. Communist Party wishes were given more weight than evidence that Komarov would not return alive. "Preliminary launches," wrote Yesikov, "showed faults in the coordination, thermal control, and parachute systems."[28] According to official accounts, it was a failure of the parachute system that killed Komarov. *Challenger*'s final launch had been rescheduled several times. Representatives of the media were tired of waiting and were making disparaging comments about NASA. This, too, may have encouraged launch approval.

Although organizations may claim that safety is their *top* priority, it is not their *only* priority. Other priorities include getting by with limited person-power and funding, simplicity of design and ease of construction, ease of operation, aesthetics, making demonstrable progress, and ensuring that the finished system is acceptable to various stakeholders. The space shuttle's design was not necessarily NASA's first choice but a compromise to keep crews in space by winning air force support by meeting military specifications. NASA did not have enough money to build the shuttle fleet properly and maintain the promised schedule of flights. Insufficient funding has many implications for safety, ranging from poor designs to antiquated equipment, sloppy maintenance, and attempts to get by with quick fixes when major overhauls are required.

Designs

We must insist on high levels of quality and reliability to increase the safety of the rockets, habitats, and other implements for exploring space. Spacefarers deserve tools that work right, do not require excessive maintenance, and do not break down. We can incorporate safety devices. Crew should be warned of impending dangers as soon as possible, and they should have multiple ways to extricate themselves. We should make sure that equipment is user friendly: easy to understand and easy to operate. Equipment should work in ways that seem right and natural, and not counter to instinct or habit. Even in the case of highly automated systems, operators should be able to reinsert themselves into the loop. They must be free to reassert control over their computers and machines when they feel they need to do so.

Quality and Reliability

One design strategy is to overengineer systems so that they will perform well even if conditions are more rugged than anticipated. Designers can choose materials that exceed minimum specifications and follow procedures that surpass industry standards. Thus, air frames may be of thicker construction and have more welds than seem necessary, or electronic circuits may be made of gold or silver rather than copper or aluminum to improve conductivity. It's possible to overengineer a spacecraft or habitat, making it able to withstand greater forces than planned, but there are penalties in the form of development cost and, in most cases, the amount of extra weight that must be carried into orbit.

NASA has always placed great faith in redundancy. Redundancy simply means providing more than one way for getting the job done. A simple example would be two independent systems intended to warn you of increasing temperature in your automobile radiator. One might consist of a thermometer hooked up to a gauge, and the other system, a pressure sensor that would turn on a light when the water pressure rises due to boiling. If one of the systems (the thermometer) failed, the other system (the pressure gauge) would still work. Give the spacecraft five computers; if four fail, then the fifth can take over. Redundancy, like overengineering, also contributes to weight and cost, and the question becomes, how many redundant systems can one afford? If four computers fail, why should we have confidence in the fifth?

NASA rightly insists upon components of the highest quality. The contractors and subcontractors are given very high standards, and their work is carefully inspected. Prior to liftoff, systems are continuously checked and monitored. Despite the high standards and constant checking, components are not always made and assembled in acceptable ways. Checks of Mercury spacecraft and, after the Apollo 1 fire, the Apollo spacecraft revealed hundreds of flaws in each. Overengineering, redundancy and quality control help, but they do not guarantee flawless performance.

Safety Devices

Systems can be engineered to include warning and fail-safe devices. The problem is that these devices can add to the mental workload. This occurred when, within minutes of the start of the Three Mile Island nuclear meltdown, scores of lights and buzzers were activated, each demanding an operator's attention. To reduce the confusion, designs can incorporate automated devices that take over in dangerous situations. Sometimes, however, such devices can actually *increase* the danger by setting into motion an invariant sequence of events that is undesirable given the specifics of the emergency. We see this in science fiction films when someone pushes the wrong button and a synthesized voice cheerfully intones, "Ten minutes to destruction."

Early (1960s) spacecraft had abort capabilities that allowed the spacefarers to separate their container from the rocket in case the rocket started to tip over or blow up. This consisted of sort of a super "ejection chair"; a small explosion could blow the capsule free of the rocket and then deploy a parachute that would safely land (everyone hoped) the people inside it.

During the shuttle's first flights, the commander and pilot had ejection seats. Later these were eliminated, in part because they seemed superfluous and in part because they could extricate only the pilots, leaving the rest of the crew to a more certain fate. Following *Challenger*, shuttles were equipped with a new escape device: after opening the shuttle's door, astronauts could maneuver along a skinny rod until they were free of the craft and then parachute to Earth. The International Space Station will be equipped with Soyuz craft or a small, one-way vehicle that will allow crewmembers to return to Earth on an emergency basis.

Safety equipment itself can pose problems. In space, such equipment displaces scientific and other equipment that makes spacefarers productive. Spacefarers must spend time undergoing drills or lose proficiency in safety equipment use. The equipment itself requires attention. For example, when an oxygen canister erupted in fire, Mir spacefarers found that some fire extinguishers were so tightly strapped to the wall that they couldn't be moved, and that some others were empty.[29]

In industrial settings, workers tend to ignore safety in the interests of expediency (it's much faster to "just do it" than to hunt for gloves and goggles). A desire to maintain a particular self-image is another motive for steering clear of safety equipment. An example is people who brave pain by eschewing the use of hearing protectors in the presence of very loud equipment, such as jet exhausts. Although they appear "braver" than people who cover their ears, they also lose some of their hearing. Cosmonauts have remained on Mir despite equipment failures that were supposed to trigger evacuation; understandably, nobody wanted to be the cosmonaut who abandoned Russia's space station. It was a tough call: the crews succeeded, and Mir remained aloft.[30]

User-Friendly Designs

Over the years, designers have followed certain conventions. The up position for a toggle switch is "on" and the down position is "off." Twisting a rotary switch or dial to the right increases the setting, while twisting it to the left decreases the setting. Simple examples here are the volume control on a radio and the channel selector on a television set. This relationship between direction and intensity carries over to displays. On a gauge, the higher values, such as a full gas tank or a high rate of speed, are to the operator's right and the lower values are to his or her left. Because people have a lifetime of experience with these conventions, it is easy for them to read gauges correctly and properly execute intended actions.

Closely related to conventions is standardization across units. If we want people to make a successful transition from one setting to another, the two settings should be as similar as possible. In shuttle cockpits, for example, we want displays and controls to be of the same type and in the same location, so that appropriate action in one shuttle is equally appropriate in another. Unfortunately, it is more difficult to achieve standardization in new and evolving advanced technological systems like spacecraft than in established, simple, and mass-produced systems such as automobiles. This is especially true when we are dealing with hundreds of contractors from many different countries.

Function should take precedence over appearance. Designers should resist the temptation to seek accolades from their peers because their product is highly aesthetic and seek instead to make displays and controls identifiable and simple to work. The neat alignment of a dozen identical toggle switches looks great in an advertisement, but it is safer to make key controls visible and accessible. In the past, nuclear power plant operators pasted notes on important controls, or even hung empty coffee cans on them, so that they could be found in a hurry. This customization made them stand out against the banks and rows of similar displays and controls.[31]

Many other design techniques ease the burden on the operator, as shown in the design of the first crewed Russian spacecraft. One of the main challenges of survival is ensuring that the spacecraft does not incinerate as it reenters Earth's atmosphere. This requires special material to withstand the ferocious temperatures, or in the case of early spacecraft, a special layer of ablative material that slowly burns away and in the process protects the hull. The first Russian space capsules were spherical. The craft was weighted in such a way that gravity ensured that the insulated section pointed downward. It thus bore the brunt of the forces that acted upon the craft when it reentered the atmosphere. The first U.S. craft, in contrast, had to be carefully oriented by means of firing small rockets. This required the highest level of attentiveness and skill on the part of the operator and increased the latitude for human error. However, the cosmonauts paid a price for their convenience because the spherical Russian configuration built up greater speed on reentry, subjecting the occupant to far greater G-forces.

Finally, operation of spacecraft should not be predicated on superhuman performance by their operators. Donald Norman points out that human behaviors are only approximations of desired behaviors and that mistakes are both likely and natural.[32] If a piece of equipment fails because an operator is tired, stressed, bored, or trying to do too many things at once, it is not designed to meet normal human specifications.

Keeping the Operator in the Loop

The central thesis in Charles Perrow's *Normal Accidents* is that many of today's systems are so complex they can experience multiple failures.[33] Various components of the DEPOSE model can fail simultaneously or in sequences that yield unpredictable and dangerous results. Two characteristics distinguish the most dangerous systems. One is parallel processing; that is, they do many things simultaneously rather than sequentially. The other is close coupling. Events rapidly follow on the heels of one another. Parallel processing allows several things to go wrong at once, while close coupling—the lack of slack or a buffer in the system—leaves little time to recover.

Spacecraft are among the many systems characterized by parallel processing and close coupling. Perrow points out that Apollo 13 is a good example of how multiple problems can arise, and how operators can extricate themselves if they are granted the opportunity to do so.

On Apollo 13, the flight that *almost* failed, a high-voltage power supply fused a switch and shut off a liquid oxygen tank. This occurred during a test that appeared to be successful, and for this reason the problem was unidentified. A caution light went on during the flight, but this was impossible to interpret because it was a multipurpose light that indicated malfunctions in both the oxygen and the hydrogen tanks. The astronauts could not examine the tank, because it was in an inaccessible location. Astronauts heard the explosion, and it was even noticed by an amateur astronomer on Earth, but it was not correctly interpreted. Both propulsion and life support systems were put at risk.

This was a series of failures that were not disastrous individually, but that combined in such a way as to seriously threaten the astronauts' survival. Not only did the switch fuse, the various warning lights and gauges failed to yield a correct diagnosis, and the tanks could not be inspected because they were housed in an inaccessible service module. Failures occurred in ways that statisticians would describe as interaction effects. Who could imagine this unlikely combination of things going wrong? Proliferating or cascading errors are annoying when they make it temporarily impossible to print out the final copy of a report, or slow down an assembly line, but are catastrophic when they occur in chemical plants, nuclear power stations, aircraft, and spaceships.

The former NASA flight director Gene Kranz described the electric response at mission control.[34] It took a few minutes to assemble the data, interpret it, and draw the conclusion that not only would there be no landing on the Moon, the crew (which was already four-fifths of the way to the Moon) would be lucky to escape with their lives. The explosion had reduced the amount of water available to the crew and robbed them of oxygen that they needed to breathe and precious fuel needed to keep the temperature of the craft at an acceptable level. Would the remaining resources be sufficient to keep them alive for the return to Earth?

Flight controllers and fellow astronauts swung into high gear, working on everything from devising and evaluating grand strategies to bring the spacecraft back to Earth to solving countless small problems that threatened the welfare of the crew. Brainstorming and improvisation were the order of the day. For example, the lunar landing module was recast as a safe haven and the crew relocated there until it was time to return to the command module to reenter the Earth's atmosphere. As engineers proposed emergency equipment, such as an emergency ventilation system for the safe haven, other support personnel attempted to build the device using cardboard, duct tape, and other supplies available on the craft. As support crew developed emergency procedures, astronauts entered Apollo simulators to try them out. Getting the tired, cold Apollo 13 crew home before they froze or their water and air ran out required some very tough calls, and support personnel kept rethinking the issues and adjusting the procedures almost to splashdown. The solution, in other words, was abandoning bureaucratic procedures and relying on human ingenuity and resourcefulness. In the case of Apollo 13, both crew and support personnel set aside standard procedures and checklists, viewed the mission in new ways, and improvised new tools and procedures that brought Lovell, Swigert, and Haise back to Earth.

Conclusion

In contemporary society, we are protected by a vast network of safety devices. Unless we live in remote rural or wilderness areas, we are easily reached by police and fire personnel, ambulance drivers, and other safety workers trained to handle emergencies and whisk us off, by helicopter if need be, to a hospital or other place for repair. Two-way radios and, increasingly, cellular telephones allow us to call for help. As we go to direct satellite services it will be possible to call 911 (or its equivalent) from any place on Earth.

No matter how much backup mission control provides, spacefarers must be prepared to survive on their own. Apart from offering advice and encouragement, there is little that outsiders can do. For example, the initial Mars voyagers will be on their own, and even if a standby rescue mission were available to go to the Moon, it would take a matter of days (perhaps weeks) to get it there. Consequently, it is essential to make space exploration systems as safe as possible and provide spacefarers with the means to cope with the accidents that do occur.

We can minimize risk but we cannot eliminate it. There have been deaths in space before, and there will be deaths again. The consequences of disasters are like the results of throwing a rock into the water. In addition to the splash of the disaster itself, there is a series of concentric rings or ripples moving outward from the center, perturbing greater and greater surface area but with lesser and lesser effect. As the immediate victims, it is of course the spacefarers themselves who will be most heavily traumatized. Those who aren't killed or physically incapacitated will bear psychological scars. These can include diminished capacity, symptoms of posttraumatic stress disorder, and feelings of guilt for having survived while their friends died. Those who survive a disaster when others do not will have to carry on although traumatized and understaffed.

Then, there are the effects on society at large. According to Leonard David, people recognized the dangers of spaceflight and were willing to accept *Challenger*'s loss until the media generated negative attitudes and public sentiment that delayed the program for years.[35] Why is it that society as a whole can overlook tens of thousands of highway deaths annually or battle casualties in foreign countries at war but get really upset by the death of a handful of spacefarers? Part of the answer is that whereas we may feel bad about the crash of a jumbo jet we can't imagine living without air transportation, but space exploration still seems optional. Part of the answer is that airplane crashes usually involve anonymous individuals, while astronauts are known individually to the public. They are not simply 1, 2, and 3, they are Dick, Ellison, and Christa. In a sense, the media bonds us with certain astronauts, and when they die we feel as if we actually knew them, and then we experience personal loss. Yet unless we are willing to accept human loss, we cannot establish ourselves in space.

CHAPTER 11

OFF DUTY

Apparently not every space crew has been dominated by relentless work schedules. The first Mir crews awoke about 8 A.M. Moscow time. Until about 10 A.M. they ate breakfast and took care of their personal hygiene needs. Next came three hours of work and one hour of vigorous exercise. After an hour lunch break, there was another four-hour block of work and exercise. Around 7 P.M. they began preparing dinner. After they ate, cosmonauts could relax, have fun, or (because not even spacefarers are exempt) catch up on their paperwork. Each week, they had two days off. There was no particular rush to start new projects if it made sense to postpone them for another day.[1] Certainly work predominated, but for the first Mir crews life was a little less frenzied than it was for participants in many other missions.

Life is divided into three spheres: work, self-maintenance, and recreation. Despite the centrality of work, spacefarers spend off-duty hours looking after themselves and perhaps taking part in recreational activities. In between missions, they must readapt to their families and cope with the public. At the end of their careers, they will retire from spacefaring and, in many cases, find new lines of work.

In space, life proceeds with a strange combination of high-tech wonders and simplicity. Marshall Savage points out that space settlers have to get by with fewer possessions.[2] They may be required to share items that aren't used on a regular basis and eliminate redundancy in their lives. This may be very difficult for people from a consumption-oriented society, where many of us own more than one telephone, television, computer, and wristwatch, and we replace such items regularly as improved versions become available.

Self-Maintenance

Self-maintenance encompasses those activities necessary to keep us alive and functioning. It includes personal hygiene, eating and drinking, and sleeping. In the long run, it must also include sexual gratification and procreation.

Personal Hygiene

Personal hygiene is complicated by cramped conditions, limited availability of water, and microgravity. As noted in the discussion of life support (see chapter 4), despite gradual improvement over time, space showers and toilets are limited compared to those that we are used to on Earth. Whether one is showering, brushing one's teeth, or shaving, there is always the problem of collecting the water, spittle, or whiskers so that they don't float around the cabin annoying fellow passengers, clogging air filters, and causing electrical shorts. We are far beyond the era of wipettes, but maintaining personal hygiene tends to be time consuming and/or to require special equipment, such as a special vacuum to suck up blobs of water after a shower.

Eating and Drinking

Food does much more than keep us alive. We prepare fancy meals that are pleasing to the eye as well as the palate, we use food as the centerpiece for a party, and we celebrate cultural differences by sampling ethnic cuisine. People in Antarctica, under the sea, and in outer space use meals to mark special occasions, including holidays (such as Independence Day and Christmas) and birthdays. In Antarctica, cooks have shown tremendous resourcefulness using canned rations to prepare banquets, complete with written menus. In space, the arrival of a newcomer to Mir is marked by ceremonial consumption of bread and onions. Newcomers may bring specialty foods from their home cultures, and spacefarers from different places may take turns trying to create regional specialties.

Some spacefarers take little or no pleasure in eating and consider meals a necessary but irritating interruption. Yet many isolated and confined people become obsessed with food quality. Good food is one of the inducements that attract people to work on Alaska's North Slope, on superships, and in submarines. High-calorie snacks are available throughout the day, and oftentimes the meals themselves feature such delicacies as filet steaks, crab legs, and lobsters. Every now and then—for example, when a French *spationaute* arrives—spacefarers have enjoyed sumptuous fare. According to Anthony R. Curtis, the "best ever space meal" was served by the French *spationaute* Jean-Loup Chrétien to five cosmonauts aboard Mir in 1988.[3] It included "compote of pigeon with dates and dried raisins, duck with artichoke, oxtail fondue with tomatoes and pickles, beef bourguignon, sauté de veau Marengo, ham and fruit pates, bread, rolls, cheeses, nuts, coffee and chocolate bars."

Current technology allows people to eat when they please, but there is an advantage to encouraging crewmembers to eat together. Food preparation and cleanup as well as the meal itself are occasions for relaxed conversation, and they promote group solidarity. Because food serves so many important social functions, Jack Stuster urges spacefarers to take the time to eat together.[4]

Alcohol is the most widely used recreational drug, and unless it causes serious problems within the family or at work we tend to ignore its effects. Users find it relaxing and a helpful "social lubricant." For some people, the thought of a beer or a glass of wine at the end of the day is a powerful motivator. On superships, at least in the early 1970s, sailors looked forward to "pour out," an evening happy hour when alcohol was made available in controlled quantities.[5] Early explorers typically stockpiled copious quantities of beer, wine, cognac, and other alcohol before departing for polar regions; in Antarctica, drinking remains a controversial but popular pastime.

Some observers wonder if wine accompanied the meals prepared by the French *spationaute,* and if vodka is delivered to Mir cosmonauts along with videotapes, books, chocolate bars, and other morale boosters contained in the "psychological support packages." Jerry Linenger reports that Mir had a well-stocked liquor cabinet and that alcohol was not for medical use only.[6] Indeed, he discovered someone had stashed a bottle of whiskey and a bottle of cognac in the arms of his spacesuit.

Unfortunately, whereas alcohol may make people claim they feel better, it also depresses performance. Heavy drinking, especially, is implicated in slow and faulty work, accidents, confrontations and fights, and almost every kind of self-destructive and criminal behavior. Drinking was implicated in the recent episode that involved two male participants in an ISS simulation mission first getting into a brawl and then making unwelcome amorous advances on a female crewmember. Whereas Linenger respected the Russians' right to drink on Mir, as a physician who is well aware of the effects of drinking he personally followed NASA's no-alcohol policy. If we get to a stage where astronauts in flight drink alcohol openly, we would expect them to consume no more than limited quantities under highly controlled conditions.

Smoking is prohibited in part because it would overtax the air purification system. At least one nicotine addict has complained that giving up smokes was one of the more demanding requirements of spaceflight. If there is a rule of thumb, it is, give up smoking in advance of the flight. Delicate docking maneuvers are not a good time for symptoms of nicotine withdrawal.

Given the heroic stature of spacefarers, drug abuse seems unimaginable, but there are several ways that this could come about. A spacefarer could become overly reliant on stimulants or sleeping pills. Someone might steal drugs from the medical kit or sneak illegal substances onboard. Drugs might be synthesized in an orbiting laboratory, and who knows what types of interesting vegetation could be found at interstellar destinations? Part of the problem of coming to grips with intoxication in space is that chemical substances do not have identical physiological consequences on Earth and in microgravity, and in the absence of experimental data we can only guess how spacefarers will react to mind-altering chemicals.

Sleeping

Sleep complaints are common in isolated and confined environments. The most common sleep complaints include insomnia, waking up in the middle of the night, and other problems that make it difficult to get enough sleep.[7] In space, sleep is disrupted by the excitement of the flight, a felt need to finish work that was not completed as scheduled during the day, voluntary activities such as writing postcards to be delivered to friends following return to Earth, high ambient noise levels, rapid shifts from daylight to darkness and back to daylight again, and having people a few inches away who are trying to work. Sometimes spacefarers are able to work their way through an incredibly demanding day, operating primarily on "nerve," and then "crash" into an exceptionally deep slumber.

On the shuttle, sleep disturbances remain a problem. Disturbances are most pronounced during the beginning and end of the mission, and depending on the astronauts' work shifts, up to 50 percent of the astronauts relied to some extent on sleeping pills.[8] Individual astronauts test types of sleeping pills and dosages prior to departure; the idea is to help them get a good night's sleep but awake refreshed, not dazed or in some sort of hypnotic state. Sleep problems are intense only during the first few days of a mission, so it should not be necessary to rely on sleeping pills throughout a long stay on the ISS.

On early U.S. missions through Apollo 9, pairs or trios of astronauts worked and rested on different shifts. In this cheek-by-jowl environment, astronauts who were trying to sleep were kept awake by companions who were trying to work. This conflict was solved by having all of the astronauts sleep and work on the same cycle, which coincided with the work-rest cycles of mission control. In larger spacecraft, such as the ISS, separate living and working quarters make it possible for people to work in shifts without disturbing companions who are trying to sleep. The Spacelab–space shuttle missions scheduled work twenty-four hours a day. This was fine because the research was going on in the laboratory while the off-duty crewmembers slept in the shuttle.

Sex in Space

Some people voluntarily choose celibacy, but we cannot expect this of everyone who enters space. Tourists, especially those on their honeymoon, will be drawn to space to experiment with sex under conditions of microgravity. Sex is a normal part of life, and spacefarers on long-term missions will seek some form of sexual gratification. Space settlements will draw entire families into space, and unless we are willing to content ourselves with test-tube babies, sex will be essential to replenish crews on multigeneration missions.

NASA has avoided few topics as studiously as the subject of sex in space. Given the selection of "right stuff" male astronauts, the brevity of the missions, and the close monitoring of life aboard the spacecraft, sex was not much of an issue during the early days. In later years, public pressures may have contributed to NASA's avoidance of the topic. NASA's approach to congressional support and funding rests in part on not annoying any appreciable segment of the population. Since sex outside of marriage (or even within marriage but at taxpayer expense) still runs against the grain of some Americans, NASA's avoidance of the topic is understandable.

Spaceflight conditions will affect the sheer mechanics of sex. Microgravity invites experimentation with previously impossible positions and acts. However, spaceflight also makes sex physically difficult and, by some North American standards, unappetizing.[9] There is little or no privacy. Lovers cannot count on gravity to stay in place—a consideration that led one inventor to develop a special leather harness that anchors one partner by the hips while nonetheless permitting undulating motions. Sweat does not collect as it would under normal gravity; rather, it forms liquid spheres that may break loose and float around the cabin. Air filtration systems are imperfect and personal hygiene facilities are limited, meaning that it is not so easy to clean up afterward. Of course, as people who have had sex in the backseat of a VW bug or in the boiler room of a tramp steamer know, none of this is prohibitive. It's just that for now, sex, like almost every other activity, will proceed without the comfort and amenities we are used to on Earth.

When we look beyond real or imagined public relations debacles and the novelty of sexual experimentation, we find profound issues of intimacy and interpersonal dynamics.[10] Spacefarers live in close confinement, and we want them to be cordial, indeed friendly, with one another. Yet we might be wary of unusually strong attachments or emotional bonds. We must count on crewmembers to work as a team and not show favoritism by attending to a lover rather than to the job. It could be very difficult to manage a personal relationship that goes sour early in a mission. After all, there is no place to escape the broken relationship, and a substitute partner could be very difficult to find. And, as is always the case during the long-term separation of partners, extramarital affairs can undermine preexisting marriages. Thus, spaceflight conditions can complicate romantic relationships that are already complicated enough.

One possibility is to compose the crew of preformed couples and hope that the different sets of partners will remain content with one another until the mission is over, and that favoritism will not get out of hand. Occasionally, someone suggests an overtly homosexual crew. This, of course, would do nothing to minimize rivalries and conflicts onboard but would do much to terrorize NASA public relations experts. Maybe the wisest course is simply not to ask and to leave spacefarers in charge of their own lives. In some spaceflight-analogous set-tings, confinees have secret, informal "provisional marriages" that last until the mission is done, at which point they terminate the relationship and return home with feigned innocence to their husbands and wives.

LEISURE TIME ACTIVITIES

Apart from working and trying to sleep, there isn't all that much to do in space. Even as early as Gemini and Apollo, astronauts reported oc-casional boredom. At this point in the history of spaceflight, there are not a lot of recreational opportunities. Spacefarers exercise, work, look out the window, then exercise again.

Like everyone else, spacefarers need ample opportunity to kick back, relax, and do nothing. Although it may seem that time given to self-contemplation and reverie is wasted, it helps spacefarers unwind, collect their thoughts, and prepare for upcoming events. Some down-time is particularly valuable after a full day of activities, since an hour or so of unwinding makes it easier to go to sleep.[11]

Self-Improvement

People in isolated and confined environments may try to improve them-selves. One option is to improve their physical condition. Exercise has recreational as well as fitness aspects, and it may be worthwhile to make exercise more fun. One way of increasing involvement is through giving the exerciser feedback (for example, load level, time spent, imag-inary distances traveled, calories burned). Another is letting exercisers earn credits that are later traded for some kind of special privilege or treat. Exercise equipment could be located near a shuttered window; operating the equipment would open the shutter and make it possible to look outside. Video game technology could be applied so that a person riding an exercise bicycle could view interesting virtual scenery, or for that matter, be a virtual contestant in the Tour de France.

Especially among scientists, reading is a popular recreational activ-ity in isolated and confined environments. Oftentimes, isolated people expect to read educational material such as science books but end up reading westerns, romance novels, and other time fillers. One way to encourage mind-expanding reading is to offer future spacefarers op-portunities to build their professional credentials in flight, perhaps by completing the equivalent of a correspondence course. There will be no need to pass tattered paperback books from one crewmember to another. Given the rapid growth of computer power, huge onboard databases could contain the same information as the Library of Congress. Specific books could be downloaded to individual spacefarers' portable computers. The electronic library itself could be periodically refreshed by uplinks carrying new titles from Earth.

For some spacefarers, self-contemplation and religious studies may be valuable leisure-time activities, since spirituality can reduce stress and strengthen fortitude. In a study of women on an Antarctic trek, Gloria Leon and her colleagues found that spirituality contributed to a sense of resolve that kept them going despite unexpected hardships.[12] Jack Stuster urges us to encourage religion and spirituality but says these activities should be private.[13] His reasoning is that tomorrow's space crews will come from many different sects, and the open practice of one religion could have a divisive effect. Religious services might become appropriate when migration to the stars gives rise to new religions that reflect the needs and belief systems of emigrants. In addition to offering a welcome change of pace, ceremonies and rituals sharpen group identity and strengthen group solidarity. Nonreligious ceremonies intended to bring people together, notes Stuster, could mark significant accomplishments (for example, a spacefarer's first trip beyond the orbit of the Moon) and important dates (such as the launching of the first cosmonaut).

Recreation

From the Mercury mission on, astronauts have demanded viewports. Some astronauts are almost transfigured by the view of Earth from space (see chapter 1). On Skylab, on the space shuttle, and on Russian space stations alike, spacefarers have clustered around viewports. This pastime could diminish on missions where long periods of time are spent in the interplanetary void, although a compact but powerful telescope might generate interest by allowing novel and unparalleled views of the heavens. Similarly, since novelty tends to wear off, long-term residents of Mars may be less enthralled by the vistas than newcomers to the planet.

Television broadcasts from Earth and videotaped movies cheered early Salyut cosmonauts and have been important ever since. Mary Connors points out that such media are essential for letting spacefarers stay abreast of the news, view baseball games and other athletic contests, and maintain ties with people and events on Earth.[14] In most cases, it will be possible to transmit news and entertainment to the space habitat, where it can be stored and played at leisure.

Getting together to view a movie—even to poke fun at it—is an enjoyable group activity that reduces tension and builds camaraderie. In Antarctica, certain films attracted dedicated cults of viewers who watched them repeatedly, sometimes chanting the actors' lines in unison.[15] However, since not everyone likes the same kind of entertainment, it makes sense to cater to individual preferences. One solution is to give everyone his or her own entertainment unit, perhaps individual cassette players or laptop computers that can be used as personal photo albums and for viewing movies when not used for work.

Isolated and confined groups sometimes shy away from competitive games such as darts or cards because these can increase social tensions in an already tense situation.[16] The winner may alienate losers, particularly if the pattern of winning is highly repetitive. Murder may be less likely if spacefarers play against outsiders, such as when cosmonauts avoid in-group competition by playing chess against people on the ground. Or perhaps competitive needs could be satisfied by playing against oneself (for example, trying to earn progressively higher scores on a computer game). Another alternative is cooperative games such as magnetized jigsaw puzzles.

Future spacefarers may find delight in simple recreational activities that are evocative of an earlier era. People on the frontier tend to be good at improvising. Studies of confinees in fallout shelters in the 1960s demonstrated the resourcefulness of isolated groups.[17] These groups, which consisted of men, women, and children of all ages, put on shows, told stories, and sang in unison. Typically, this entertainment was organized by women and intended to amuse the children.

Maintaining Contact with Family and Friends

Communication with family and friends is an important off-duty activity for people who are physically isolated from their families. Early explorers looked forward to mail, newspapers, and shreds of information from their hometowns. Contemporary submariners must maintain radio silence but on rare occasions receive brief messages from home. We know also that people in Antarctica look forward to talking with relatives and, in the days before good satellite communications

were available, were distressed when weather conditions or other factors made communication impossible. On Mir, limited communication
with the ground (caused by a paucity of Russian monitoring stations)
was one of the hardships. On the ISS it will be somewhat easier to
provide spacefarers with good private communication with their family and friends on Earth.

We know from studies in Antarctica and elsewhere that communication with home is a mixed blessing. People who receive little mail
may become depressed, yet isolation and confinement may seem less
tolerable when one is reminded of all of the wonderful things that have
been left behind. Isolated people may become anxious if they receive
the equivalent of a busy signal or if nobody is home to accept the
transmission. Such apprehensions are heightened if there is a long wait
before the next opportunity to call. Assuming that the person at home
is not deliberately snubbing the spacefarer, this problem is overcome
by advance scheduling.

Not all words from home are comforting. Spacefarers' families are
like other families, with their occasional marital conflicts, problems to
solve, and bills to pay. Spacefarers may be disconcerted by reports that
a relative is ill, that a son or daughter is having trouble in school, or
that the repossession agency is eyeing the new Corvette. Isolation and
confinement put the relationship under stress, and proof of this may
emerge in a conversation with home. Thus, the reward of "phoning
home" may be a squabble or fight, word that a fiancée has decided to
break off the relationship, or broad hints that a divorce is imminent.
A spacefarer may feel helpless because he or she is too far away to take
corrective action.

It is particularly difficult if there is really terrible news, such as the
death of a parent. Such news, and the resulting level of emotional dis-
tress, could jeopardize performance. The issue of whether or not to
relay bad news arose when a cosmonaut's father died of a heart attack.[18] Choosing not to inform the cosmonaut, people on the ground
found plausible, innocuous reasons why his father could not come to
the telephone and were able to deflect the cosmonaut's questions without arousing suspicion. Spacefarers interviewed on this topic are about
evenly divided: half would like to be informed, whereas the other half
would not. Although the choice of whether or not to receive bad news
could be an individual decision, one wonders if those who elect *not* to
be informed would worry that important news was being withheld
from them.

Communicating by radio or other electronic media is not the same as communicating face-to-face, and different communications media— e-mail, voice, video—affect the emotional tone of the interaction as well as people's understanding of the message.[19] Media differ in terms of their "richness," or the amount of information they convey. Compared to reading e-mail, listening to someone speak, even over the radio, gives us an idea about that person's emotional state. Rate and amplitude of speech, tenor of voice, hesitations, and pauses are highly informative. We learn even more when we can see that person (even if only on a TV screen), because his or her words are supplemented by postures, gestures, and facial expressions.

For impersonal, businesslike communications an e-mail message will do. When it comes to communicating with family and friends, or receiving congratulations from important people, most people prefer the information-rich media communications systems. Thanks to rapid advances in information compression (which makes it possible to squeeze a lot of information into a thin slice of a radio band), interactive video is available on some of today's missions and may be standard in the future.

Radio waves travel at the speed of light. This means that for most purposes radio communication is instantaneous. Delays of just a few seconds can make the other person seem dimwitted (by seeming slow to answer your question) or lacking a sense of humor (because so much time passes between joke and laughter). Conversations become awkward and it is easy for the participants to lose synchronization.

Communication from people on satellites in near-Earth orbit are no problem—after all, as spacefarers hurtle by Houston or Moscow they may be only a couple of hundred miles from the closest radio station on Earth. It is easy, also, to converse with people on the Moon, although there are noticeable pauses. However, as we consider expeditions to Mars, delays will range from six to forty minutes round-trip, making normal conversation impossible. At Mars and beyond we may shift from conversations to sequential monologues. On eventual interstellar missions, communications will become one-way. At some point spacefarers might forget their question before the answer arrives. Here we can learn from a standard practice in e-mail: people on Earth can return the spacefarers' original question when they send their reply.

DOWN TO EARTH

Until the time that people set forth on lifetime (one-way) missions, spacefarers will return home, temporarily between missions and permanently for retirement. As Philip Harris points out, when we consider the psychological and social aspects of spaceflight we must include the spacefarer's relationship to his or her family, home community, and society.[20]

Family Relationships

Many worthwhile activities require people to be away from their families for extended periods of time. These include overseas military assignments, seafaring, work shifts on Alaska's North Slope, and taking part in extended spaceflight. Unlike many people who work in remote settings, spacefarers on the ISS or Mars will have no breaks or vacations that would allow them to visit home partway through their missions.

Primarily from studies of military personnel, we know some of the frictions associated with separation and reunification.[21] The partner who remains at home must shoulder the entire burden of managing the house and raising the children. Any emergency or problem must be handled by this person. The spacefarer's spouse knows that spacefaring is an extremely dangerous line of work and may be particularly concerned during launch, reentry, or other critical parts of a mission, and absolutely terrified when it appears (as it has on several occasions) that the spacefarer has been lost. Even during a "normal" mission, the partner at home may feel as if he or she were living in a fishbowl, continually under the scrutiny of well-meaning but intrusive people and besieged by the media. Outsiders' attention may be especially intense during the most unsettling parts of the mission.

Spacefarers miss important occasions—a new job or promotion for the partner at home, weddings, funerals, graduations, anniversaries, and birthdays, perhaps even the birth of a child. The spacefarer knows that he or she is temporarily helpless when it comes to family matters. All of this, perhaps augmented by the knowledge that it is the spacefarer, not the long-suffering partner, who is the center of attention, can anger the partner and confer a sense of guilt upon the spacefarer.

The situation changes rapidly and not necessarily for the better when the wanderer returns and tries to resume the roles of spouse and parent. He or she may try to relieve the partner of tasks that the partner has learned to enjoy, or criticize the partner's ways of doing things that worked fine when the spacefarer was away. In some military families, for example, the wife takes charge of the home front but is expected to relinquish command to the warrior-husband when he returns. The ensuing tensions can destroy the romantic vision that each partner had of how wonderful it would be when they were finally reunited.

Frictions are exacerbated if the spacefarer accepts obligations that prevent full participation in home life even when on the ground. Unlike the North Slope workers who have generous vacations, or military personnel on extended furloughs, almost immediately following their return some spacefarers make public appearances, meet with dignitaries, and start training for their next mission. If the spacefarer has workaholic tendencies anyway, it may be very difficult to become reengaged in family life even when there is explicit time off to do so.

Infidelity is rampant in our society, and why should we expect spacefarers to be exceptions? Spacefaring is a glamorous activity, and since spacefarers meet large numbers of people anyway, it may be very easy for them to find attractive members of the opposite sex. Deke Slayton mentions that one astronaut was disqualified by the prospects of a messy divorce, while other marriages were glued together for the sake of a mission, but that in the final analysis many astronauts have been divorced.[22]

Working with the Public

Spacefarers are celebrities who are highly regarded by the public. Personal and media appearances can be seductive. First, acclaim is flattering and, if not overdone, rewarding. Second, public appearances give spacefarers opportunities to promote their profession and build goodwill for NASA. They can do this through educating the public and building grassroots support.

Nationally and internationally, the earliest spacefarers received the highest levels of acclaim. Rewards included public appearances, ticker tape parades, and visits to aerospace factories to receive cheers from the workers who made their missions possible. Many of these spacefarers were welcomed in international circles as they carried the prestige of their home countries to foreign lands. In a sense, the early astronauts and cosmonauts had their pick of political leaders, captains of industry, celebrities, and interesting people of all sorts. They lived in a rich and heady era.

Robert Helmreich and his colleagues believe as do many psychologists that social recognition is an important reward.[23] They point out that social recognition for spaceflight has declined since the halcyon days of the 1960s. They add that the costs of spaceflight (in terms of danger and hardship) have also declined. However, social recognition has declined at a faster rate. According to their 1980 analysis, the balance between the rewards and costs is shifting in a negative direction, making it less easy to attract and retain the best people. To offset declining social recognition, we must provide spacefarers with other rewards, such as more habitable spacecraft and greater autonomy. Yet, certain future missions could rekindle some of the public enthusiasm of the 1960s, perhaps boosting social recognition to record levels. The first crew to return from Mars will earn such acclaim. And even as there will always be military heroes, there will always be space heroes.

Spacefarers tend to be work oriented, but "working the public" is not necessarily one of the duties that attract them to their jobs. Some spacefarers, coming from strong science backgrounds, may be introverted and not all that comfortable in public. Then too there is the necessity to remain on guard, being very careful not to criticize other spacefarers or the space program. Like any celebrity on tour, spacefarers may find themselves on a fast-paced schedule, with little privacy and with a considerable drain on their time and energy.

Retirement

Eventually each spacefarer's career comes to an end as he or she makes a permanent return to Earth. For a payload specialist, this could come after one mission. For other spacefarers, this could occur any time, depending on their fitness, the requirements for upcoming missions, and the opportunities that await them in "civilian" life. So far, many astronauts have found second careers on the lecture circuit or in private industry, oftentimes with contractors in the aerospace or defense industries. This allows them to put their hard-won knowledge to good use. Others apply their advanced degrees and retrench in university or research settings. Being an astronaut is something like serving in the military: the expectation is that retirement will be followed by a second career.

For some spacefarers, leaving the limelight is welcome, but for others it is a source of frustration. No longer do they bask in the applause of the crowd. There are fewer champagne dinners with dignitaries and more backyard barbecues with friends. Marital relations that somehow survived the rigors of the spacefarer's career may destabilize and fall apart. The departure from a glamorous line of work, coupled with encroaching age, diminishes opportunities for excitement. Spacefarers are neither better nor worse off than other young retirees, but there are unique combinations of pressures that act upon them in retirement, just as there were unique pressures during their spacefaring years.

In the distant future, perhaps on Mars or beyond, retirement may mean retiring in space. This could happen when it becomes more efficient to maintain retirees in space than to return them to Earth, or when there is simply no way back. Perhaps a good example of this would be the case of a person who leaves on a multigeneration interstellar mission at age twenty, and for health or other reasons retired forty years later. It will be interesting to see what kinds of physical and cultural arrangements will evolve to support such people in comfort following long and successful spacefaring careers. Will they end their days as honored elders or as a drain on the community? In some improvident environments, people who are no longer able to contribute are expected to commit "altruistic" suicide—that is, suicide for the benefit of the group—by remaining behind or walking out into the desert or ice field to die.

CONCLUSION

Spacefarers, like everyone else, need balance in life. Despite strong interests in their fields of expertise, despite their love for space, despite workaholic tendencies and unbounded enthusiasm for the mission, there is more to life than work. This chapter took a comprehensive look at spacefarers' lives and focused on those periods when they are off duty. These include self-maintenance and leisure time in space, furloughs between flights, and retirement.

It becomes increasingly important to make ample provisions for leisure time activities as crews spend longer and longer periods aloft. Different astronauts will have different leisure time preferences, so planners must accommodate many different tastes and needs. Some pastimes are solitary and allow the spacefarer to temporarily "get off stage." Some of the pastimes, such as reading informative books or undertaking a new exercise regimen, lead to self-improvement; but the chief benefits of other pastimes, such as looking out the window, reading cheap novels, or watching popular movies, are emotional release and relaxation.

Between missions, spacefarers are united with families and friends. Although spacefarers and their families may look forward to the spacefarers' homecoming, reunions may be marred by tensions that stem from the spacefarer's absence, feelings or frustration and guilt, and the reallocation of responsibilities following the spacefarer's return. Additional difficulties may arise as spacefarers cope with the media and public.

At some point—perhaps after one or two missions, perhaps after thirty years of service—the spacefarer retires. Most spacefarers enter a second career. Here again, though, the expectations and demands of the spacefarer's role, coupled with personal experiences in space itself, may make it difficult to adapt. If too many spacefarers pay too big a price for their flights or are unable to build a new life following their retirement from space, some other spacefarers will quit, and the most qualified civilians who might otherwise be interested may not apply for spacefaring jobs.

CHAPTER 12

SPACE TOURISM

Although as a paid passenger Toyohiro Akiyama had the assignment of accumulating videotape and broadcasting live during the ten-minute periods when Mir was in television contact with Japan, David M. Harland proposes that as the first fee-paying visitor to Mir, Akiyama qualified as a space tourist. The "stereotyped Japanese tourist," Harland writes, Akiyama "brought with him half a dozen cameras and a hundred rolls of film to augment the several hundred kilograms of television equipment that had already been brought up."[1] Akiyama had trouble adapting to microgravity, experienced space sickness, and vomited. But looking back, Akiyama gave the flight high marks.[2] He described weightlessness and seeing the Earth from space as joyous experiences that many people would find attractive. Sightseeing, he wrote, was very special: the lights and colors combined with the movement of the spacecraft produced a visual feast that seemed like music.

Tourists use discretionary funds to travel and visit destinations that remove them from the routine of everyday life. They may seek adventure, rest and relaxation, or even pampering, but in all cases they are paying for experiences. Tourism is the largest industry in the world. Because it could yield benefits for NASA, the space industry, and the public, NASA and a consortium of firms known as the Space Transportation Association are working to develop space tourism.[3] Apparently, NASA has backed off from the view that space tourism is science fiction and now acknowledges its potential value as an approach to space commercialization.

In the 1960s, Krafft Ehricke and Conrad Hilton proposed visiting space for no other purpose than having fun, and thousands of people made down payments on proposed trips and reserved seats on the first spaceflights promised by a major commercial airline carrier.[4] In 1985, Society Expeditions, a leading adventure tourism company, announced intentions in this area. Today, several firms are eyeing tourist dollars. Surveys conducted in Japan, the United States, and Canada suggest that about 70 percent of the Japanese and 60 percent of the North Americans surveyed were interested in traveling to space.[5] In Japan, men and women were equally enthusiastic, but in the United States and Canada men were slightly more interested than women. As one might expect, people under forty were more likely to be interested in space travel (75 percent) than were people over sixty (25 percent). In these surveys, many interviewees indicated that they would like to visit space more than once, and some of the older respondents stated that they would have visited space if trips had been available when they were younger. A small proportion claimed they would spend more than one year's salary for the opportunity to visit space.

J. G. Pearsall points out that both the number of potential tourists and the amount of discretionary money that they have to spend will increase during the first decades of our new millennium.[6] Unlike that required for building solar-powered satellites, mining the Moon, or other commercial activities, most of the technology required for space tourism is available already.

John Spencer and the Space Tourism Society point out that if we think of tourists as buying experiences, then there are space tourists right now. His broad-based definition of space tourism encompasses not only tourist activities in space itself but also visiting Earth-based museums such as the National Air and Space Museum (which attracts millions of space tourists each year), participating in space camps, enjoying theme parks, and exploring space through virtual reality.[7] Space camps are like other camps in that they fall somewhere between a school and an amusement park. Instead of spending two weeks riding horses, rowing boats, learning how to tie knots, and eating hot dogs in a tent, space campers spend two weeks learning about the universe and spaceflight, riding in simulators, using astronaut gear, and enjoying freeze-dried ice cream. Space camps cater to both children and adults. Some companies have sent groups of employees to train together in a space camp. In the course of this, they get to know one another better and develop teamwork.

Theme parks are places like Disneyland's Tomorrowland where visitors enjoy futuristic entertainment with spacelike overtones. One Tomorrowland ride involves entering a futuristic buslike cabin that an animated robot then flies through simulated space. The craft banks and turns in response to the robot's control, while spacescapes reminiscent of those in *Star Wars* career by the windshield. Finally, today's space tourists can use computer-generated pictures and virtual reality to gain the impression of walking on the surface of the Moon or Mars or flying through an asteroid belt.

As for space itself, the big challenge is finding safe, reliable ways to get people there while keeping costs low enough so that there will be a large market. Several companies are developing vehicles to take people to the edge of space.[8] The initial cost for a very high altitude (one-hundred-kilometer) suborbital flight could be on the order of $100,000. Although this is a lot of money, keep in mind that the tourists who buy the best accommodations on a major cruise ship may pay $350,000 for a leisurely around-the-world tour.

Space tourism advocates believe that as the infrastructure is put into place, and multiple spacecraft come into use, cost will decrease to such a level that space tours can be brought within the reach of many middle-income people, at least those who are willing to save and wait for the privilege of flying in space. Furthermore, whereas tourist spacecraft may never be as uncomfortable as early spacecraft, we can expect a trend toward increased comfort and luxury. Over time, the average space vacation will become less adventurous and more luxurious. Perhaps a good analogy here is the gradual replacement of tourist camps with luxury motels as the United States' population, economy, and road system grew.

TOURIST-FRIENDLY SPACEFLIGHT

The tourist industry requires that both transportation and destination be safe, accessible, and comfortable enough to provide tourists with a positive experience. If only a few people are eligible for space, if they get hurt or sick, or if they just plain have a bad time, then word of mouth will discourage other potential customers. This may be especially true when space tours are accessible only to the rich and famous, whose descriptions of their experiences (favorable or unfavorable) will rivet the public's attention.

Who Can Go?

Only 1 or 2 percent of the applicants are accepted for the astronaut corps (see chapter 6). Once accepted, they are trained and retrained, primarily in the operation of the spacecraft's systems but also in how to live in space. Of course, space tourists will not have to fly a spacecraft any more than people touring England will have to drive a tour bus. Still, space tour operators will have to reckon with the medical risks of acceleration, microgravity, and other potentially hazardous conditions. Passengers will have to be screened and be given a level of training that goes beyond the brief safety instructions that are regurgitated at the start of a commercial airplane flight.

Presumably, space tourists will be in good cardiovascular health and free of impairments that might pose a liability in a spaceflight environment. However, the more stringent the requirements, the smaller the pool of potential tourists and the less lucrative the industry. This means establishing basic requirements for physical fitness, making the trip as easy and comfortable as possible, having medical personnel and supplies onboard, and establishing procedures for emergency evacuation. Space tourism will face an interesting challenge if people denied passage on the basis of physical criteria demand "reasonable accommodation," that is, that the spacecraft or tour be modified to take their limitations into account.

Compared to astronauts, space tourists will have to get by with very superficial levels of training. There may be a few wealthy, rugged people willing to spend months training with fellow space enthusiasts, learning how to cope with danger and stress through mountain climbing, undersea adventures, and other exercises intended to weld them into a superbly fit, high performance team. However, most people will not have the money, time, or inclination for this. Limited training, lasting a week or so, may be cast as a fun activity and worked into the vacation package. This training would focus on immediate issues, such as what to expect, how to follow procedures, and how to do things in microgravity.

Tourist Accommodations

Space tourist facilities must be engineered in such a way as to reduce risk to a minimum, even for superficially trained people. This means that the vehicles should be no more likely to fail than commercial airliners and that tourists are no more likely to get sick or hurt than on a terrestrial tour. Safety is one of the reasons most commonly cited by people who are not interested in traveling into space.[9] Customers will be reassured when insurance companies offer space travel and air travel coverage for the same rates.

Safety requires minimizing the biomedical hazards of spaceflight. Most likely, we can keep acceleration within acceptable limits. At least at first, most tourists will not proceed to the Van Allen belts, which contain trapped radiation, so in the absence of major solar flares, radiation is unlikely to be much of a problem. On the basis of medical considerations alone, under some conditions the crewmembers should be given greater protection against radiation than tourists need be given. In the case of a minor solar storm, for example, it may be the crew rather than the passengers who should crowd into the safe haven, because it is the crew, with their months or years in space, who run the risk of receiving the highest cumulative doses of radiation. No matter how practical this might be, though, passengers would find it unacceptable, so it will be necessary to include a "storm shelter" that can accommodate everyone.

Space adaptation syndrome is another matter. Note that in retrospect Akiyama enjoyed his visit to space despite vomiting. Nonetheless, sickness can diminish a sightseeing experience at the time. Simple over-the-counter pills intended to combat motion sickness will help many tourists; unlike professional spacefarers, the tourists can risk side effects such as drowsiness. Nonetheless, to encourage more people to spend more time in space, artificial gravity would help. Although this might be developed first for the tourist industry, artificial gravity would also be useful if and when people choose to live on large orbiting communities or set forth over vast interplanetary distances.

In addition to reducing the chances of space adaptation syndrome, artificial gravity would make it easy to move from place to place when the tourists are tired of experimenting with weightlessness and facilitate the preparation of and consumption of gourmet meals. The most feasible way to create artificial gravity is through centrifugation. This involves spinning people around to build up G-force. The amount of force depends on the radius of the spin and the number of revolutions per minute (rpm). Imagine yourself standing in a large coffee tin that is attached to a rope, and that is being spun around and around by a giant as if she were participating in a shot-put contest. The longer the rope, and the faster she spins, the greater the G-force or artificial gravity. If the rope is shortened she must spin faster to cause the same effect on you, but if the rope is lengthened she need not spin quite so fast. By spinning wheel-shaped or cylindrical space stations we can create artificial gravity, with the gravity being stronger as one proceeds out from the axis of rotation, or hub (see chapter 13). The hub will remain a gravity-free zone. This is important, because weightlessness will be one of the tourist attractions.

Larry Lemke points out that human subjects have lived for days in chambers that were spun around at the end of giant arms.[10] The faster the spin, the greater the motion sickness, and the longer it takes to adapt, both to the "flight" and then again after the contraption stops. There are few biomedical problems if the speed is kept at 2 rpm, but at this slow speed the radius for the centrifuge would be 223 meters to maintain 1-G.

Drawing on a study conducted at Marshall Spaceflight Center in 1987, Lemke proposes an interplanetary spacecraft that would look like a huge rotating set of dumbbells. One of the spheres, which would provide the hub for the rotation, could be used as a microgravity lab or for storing equipment and supplies. The second sphere, connected by a long tether, would swing around the first. The length of the tether and the speed of rotation could be varied so that, on a journey to Mars, for example, there would be .38-G to acclimate the spacefarers, but on the return voyage rotational speed could be adjusted to 1-G to prepare the crew for its return to Earth. A spacecraft with this capability would weigh only a little more than one without it: Lemke estimates the weight penalty at about 10–20 percent.

In addition to safety and protection from biomedical hazards, tourist facilities must achieve high levels of habitability. There are always some tourists willing to undergo hardship to visit a remote part of Earth, and many of them will choose to do the same to visit space, but space tourism as an industry will not reach maturity until the average person finds the accommodations comfortable. Tourists will require clean, private quarters, a fresh atmosphere (the scent of an old locker room won't do), sufficient if not abundant water to accommodate personal hygiene needs, interesting food, and, when the novelty wears off, entertainment. In the interests of safety and service, the ratio of staff to tourists must be high.

Tourist Activities

There are two basic models for space tourism. The adventure model, which seems a little more attainable right now, stresses scarcity and hardship, such as one might find on a mountain-climbing expedition or a trek across Antarctica. The cruise ship model stresses abundance and luxury, such as one might find on a cruise ship at sea or at a resort. John Spencer and the Space Tourism Society favor the second model and consider it necessary to stress the beautiful, sensual, and futuristic aspects of space tourism to attract financial backing and build interest among tourists.

As presently envisioned, space tourism is likely to follow the path of spacefaring itself. Suborbital flights will take people to a very high altitude to give them a good view of Earth and brief exposure to microgravity. Next would come orbital tours, reminiscent of circling Earth a few times in a space shuttle or an oversized Apollo command module. Later, there may be flights to specific destinations such as orbiting hotels or resorts on the Moon. Suborbital flights are already on the horizon, but the next steps beyond that could be big ones. Thus, this discussion is necessarily speculative.

Suborbital Flights

The first space tourists, limited to suborbital flights, will have brief but intense experiences. They will get to see Earth from an altitude that very few people attain, and they will gain limited firsthand experience with weightlessness. Moreover, as Leonard David points out, they will gain "bragging rights."[11]

If plans are realized, tourists who book trips with the Zegrahm Company will be brought to space on a ship known as a Space Cruiser, which becomes airborne via a Sky Lifter jet transport. The Space Cruiser separates from the Sky Lifter and uses its own engines to reach the edge of space and fly a large parabola, and then returns as a normal plane. Passengers booked with the Civilian Astronaut Corps eventually may ride the *Mayflower,* an ocean-launched titanium rocket.[12]

Harvey Wichman and his students have explored some of the human factors issues for tourists on suborbital flights.[13] Their report opens by tracing the hypothetical experiences of Bill and Liz Marquitz, world travelers who at age sixty are ready for their first trip into space. Their experience begins about a week before liftoff, when they arrive at the Zegrahm facilities in the Mojave Desert. In luxurious surroundings they learn about spaceflight and undergo specific instruction in safety, basic procedures, what to expect during different phases of the flight, and how to "translate," that is, move from place to place in microgravity. A simulated flight in a model of the Space Cruiser is a highlight of the training session and adds to the preflight fun and excitement.

Prior to liftoff, Bill, Liz, and four fellow tourists are issued flight suits that Wichman's students designed to serve two purposes. First, these are highly functional outfits for the flight itself. Fire-resistant and comfortable, these suits are two-piece, since one-piece outfits are much more difficult to manage in cramped locations such as a tiny lavatory. The garment is free of unnecessary straps and protuberances but contains many pockets so that people can bring small items into space. Soft slippers are used as footwear; not only are these more comfortable than hard shoes, they are less likely to cause injuries if wearers bump into each other while floating and tumbling in microgravity.

Other functions of these outfits include commemorating the experience, serving as tangible mementos, and advertising the flight to other potential customers. Attractively made, with a dark blue pullover top and light blue trousers, they are intended to last. Because the clientele will be international, a small rendition of the tourist's national flag will be sewn on the right shoulder. Other patches suggestive of those designed for specific NASA missions will also help commemorate the flight.

The Space Cruiser, reminiscent of a small corporate jet, carries a pilot and copilot along with the six passengers. Slowly the mother ship gains altitude. The first part of the ride does not differ that much from a ride in a small jet, apart from being tacked onto the Sky Lifter's back. However, after everyone makes a final trip to the lavatory and engages in a brief flurry of radio communication, the spacecraft's engine ignites and the cruiser leaves its transport behind as it gains altitude at a ferocious rate. The sky turns from blue to black, and despite the sunlight passengers get a great view of the stars—and there are more visible stars than they ever imagined.

As the Space Cruiser's trajectory shifts from a beeline to a gentle arc, the travelers experience about two minutes of weightlessness. Seats retract into the floor (to give the tourists more room to move around). They experiment with small objects that they brought with them and enjoy floating in different positions. Handholds, footholds, and bungee cords keep individual tourists within their assigned spaces, minimizing the risk of serious collisions. After a few minutes the arc is completed and gravity resumes its pull. The seats reemerge from the floor, and the passengers strap themselves back in. The spaceplane lands at the Mojave facilities; the entire flight took about two and a half hours.

Orbital Flights

Orbiting vacations would extend the period of weightlessness from minutes to hours or days and offer tourists the opportunity to observe firsthand large parts of Earth. Right now, it costs millions to put someone into orbit. D. M. Ashford of Bristol Aerospace proposes two generations of reusable spaceplanes for lifting tourists into orbit.[14] According to his analysis, these will reduce the costs of putting a person into orbit to 1 or 2 percent of what they are today. This would allow a pricing structure that would put orbital tours within the reach of tens, perhaps hundreds of millions of people worldwide.

Like the Space Cruiser, Ashford's spaceplanes would take off on the back of large, fast transports (in this case something like long, skinny Concordes) and, at an appropriate altitude and speed, would separate from the host for the next phase of the journey. The first passenger-carrying version, the Spacecab, would carry six passengers and a crew of two, but all the way into orbit. Since Ashford projects that the Spacecab would be much less expensive than a shuttle to build and operate, it could be financed in part by contracts to ferry government astronauts to space stations. After the Spacecab would come the Spacebus, a much larger version of the Spacecab, with fifty seats. Of course, this would be very expensive to develop, but Ashford explains that we need to look beyond the initial cost to some future time when the Spacebus is a mature vehicle. That is, the spacecraft will be less expensive when the up-front development costs have been swallowed and there are several dozen, perhaps scores, of each in operation. He projects that a ticket on a mature fifty-person spaceplane would cost on the order of ten thousand dollars.

Another alternative, favored by the Japanese Rocket Society, is a fully reusable rocket capable of reaching orbit with just one stage.[15] Insisting on a fully reusable rocket keeps the cost down, because there is no need to replace expensive components for each flight. Having a single stage that orbits is crucial, so that it can be launched from spaceports all over the globe. Since multistage rockets jettison spent stages, they must be launched in coastal regions, where castoffs can splash down harmlessly. Since a single-stage-to-orbit rocket does not jettison parts, it is as conveniently launched from Nevada or Kansas as from a Carolina beach.

The Japanese Rocket Society plans call for squat, blunt, bullet-shaped rockets capable of carrying fifty tourists, whose seats would be arranged in two semicircular rows facing windows. These craft would be equipped with lavatories, galleys, television sets and other amenities associated with the large aircraft, and a small area for experimenting with microgravity. There would be a beverage and food service, and of course, the opportunity to buy souvenirs. The standard Earth-observation orbit would be two hundred kilometers.

Prospective tourists would have a choice of either two-orbit or one-day vacations. Those choosing brief flights would have a choice of two itineraries. A daytime departure (from Japan) would feature good views of the Pacific and the Americas, while a nighttime departure would be dominated by views of Africa and Asia. Those on the full-day tour would get a good overview of Earth (because during each orbit they would fly over different territory) and a good taste of microgravity. According to optimistic projections, there might be as many as fifty rockets in almost continuous use, which means that, as with a commercial aircraft, there would have to be a rapid turnaround time. (Remember: a rocket on the ground is like a ship in port—it is losing money.) If desired levels of demand are realized, fifty thousand people a year would take orbital tours for about the same price as a ride on the Spacebus.

Hotels and Resorts

The next level of space tourism involves spending several days in space; for example, a week in an orbiting space hotel. Presumably, the very first orbital hotels will be reminiscent of current space stations, and there have been efforts to refurbish Mir as a tourist attraction. Ashford offers us some idea of what an early space hotel might be like.[16] He envisions a cylindrical space station that accommodates eighteen guests and a crew of six. Tourists would have private sleeping quarters. There would be an eating and meeting room, a viewing room or observatory, a microgravity lab where the tourists could conduct their own experiments, and a microgravity gym. Later, space hotels could become huge, sumptuous affairs, perhaps patterned after the best hotels and resorts on Earth. As self-contained "worlds" they could be designed to carry various motifs: for example, they could take their themes from the movies *2001: A Space Odyssey* or *Star Trek*. For the history buff, spacecraft could carry modules that reproduce earlier spacecraft such as Mir or Apollo.

Pearsall compares developing space hotels to developing the Chunnel, the tunnel that runs under the English Channel connecting England and France.[17] As with the Chunnel, developing a space hotel will require an immense planning effort and massive investment; also as was the case with the Chunnel, there is sufficient promise of economic return that it is worth committing to more detailed studies of the feasibility of such a hotel. His preliminary analysis suggests that a luxury hotel would be a great attraction but very expensive to construct, forcing vacations to be priced beyond the reach of all but the tiniest sliver of the tourist market. A Spartan hotel would be much cheaper to construct and could be made available to vacationers at a relatively attractive price, but if all it offers are tiny viewports and granola bars, how many people would want to go there? Pearsall concludes that if we assume a launch date of 2010, a modest hotel (something between a Spartan hotel and a luxury hotel) could begin to turn a profit in a decade or two.

The contemporary cruise ship, notes Robert L. Haltermann, may be a useful analogue for luxury orbiting hotels and space resorts.[18] Initially, passenger ships were used to transport people from one place to another, but by the late 1960s almost everybody traveled by air. During the 1960s and 1970s, the industry shifted from passenger service to vacation cruises. Many of today's cruise ships are larger and better, and they carry more people, than the fabled luxury ships that transported passengers from Europe to the Americas during the first half of the twentieth century. Cruises are a multi-billion-dollar industry, with companies vying to build the newest, largest, and most luxurious ships.

Cruise ships are supported by a vast infrastructure, including travel agents, food and beverage suppliers, docking facilities, maintenance crews and facilities—an infrastructure invisible to most passengers, but which (along with crew salaries, provisions, and ship depreciation) has to be factored into the cost of the operation. Most cruise ships offer a range of accommodations geared to different budgets and tastes. Although all rooms are clean and comfortable, they differ in terms of size and amenities. For example, cruise ships offer several grades of rooms, ranging from small interior cabins that lack portholes to spacious, lavishly appointed luxury suites. However, all grades of accommodations are accompanied by fine food and good service. All passengers have access to the ship's restaurants, lounges, bars, theaters, shops, and casino.

The experience of the cruise industry shows, first, that ships that were initially somewhat Spartan and used for necessity later became luxurious and were used for pleasure. Even as it would be difficult for pilgrims on the *Mayflower* to imagine today's luxury liners, it may be difficult for those of us who are familiar with today's spaceflight to envision the luxury spacecraft of tomorrow. Second, the cruise industry has identified some important cost-cutting strategies. For example, in the past, blueprints were used to construct either one or two ships. Today, shipbuilders use the same plans to construct several ships. Although this isn't mass production in the same sense as an automobile assembly line, standardization cuts costs materially. Third, the industry shows that people enjoy traveling and are willing to pay for high-quality service.

Initially, people will be drawn to space to view Earth and the Moon and to experience microgravity. However, at some point spectacular views and excellent cuisine will not be enough: they will seek additional entertainment. Special entertainment will be particularly important to attract repeat customers.

Honeymooners and couples celebrating their anniversaries may be attracted by the idea of experimenting with sex in space, so perhaps there will be some honeymoon suites. (Couples may have to settle for a shower, rather than a huge, heart-shaped tub like those found at many honeymoon resorts on Earth.) We might expect that a gymnasium would be a major attraction. Imagine jumping, tumbling, and doing somersaults unfettered by gravity, or perhaps bouncing back and forth between two trampolines. In the evenings, the gym could be converted to a theater where guests could enjoy watching live 3-D baseball and other games made possible or embellished by microgravity. By the same token, weightlessness would make possible some truly remarkable feats of acrobatics and choreography.

One of the problems faced by the hospitality industry is that it can take a lot of servers, bartenders, and house cleaners to assure top-quality service. (On a cruise ship there is likely to be one crewmember for every two guests.) Perhaps some of this work will be done by sophisticated robots. A robot bartender, for example, wouldn't require a bunk, three meals a day, pay, and periodic visits to Earth, and it could add to the tourist experience by contributing to a futuristic, high-tech ambiance.

Even more exotic than an orbiting hotel would be a spaceship that leaves Earth's orbit and circles around the Moon before returning. This would allow tourists to see Earth at a distance, view the Moon up close, enjoy an "Earthrise," and perhaps undergo a transformational experience. Perhaps the spacecraft could be equipped with powerful and securely anchored telescopes to give vacationers a good look at the stars and the lunar surface.

If space tourism matures, a resort on the Moon would be a logical destination. According to Ben Bova, visitors to the Moon will see startling effects brought about by the apparent closeness of the horizon (it will seem half as far away as it does on Earth) and, due to the lack of atmosphere, exceptionally sharp views of lunar landmarks and skies.[19] Someday it may be possible for lunar and Mars visitors to go outside the craft for recreational purposes, and Bova recommends a buddy system so that people go two at a time. Strollers will be kept to established pathways. Hopefully, this will be enforced better than it is in Antarctica, where every now and then someone tries to take a shortcut and ends up perishing after falling into a crevasse. Every kilometer or so there would be an emergency shack where people can retreat if they develop a problem with their space suit or receive a solar flare alarm. According to the former astronaut Jim Lovell, three major Japanese companies have spent a total of more than forty million dollars over the past ten years studying ways to create resorts on the Moon.[20] These studies include everything from construction techniques to developing workable golf courses.

Patrick Collins believes that many people who are now alive will live to see space tourism become an established industry.[21] He expects that by the year 2030 there will be twenty orbital sports centers, seventy-two orbital hotels, and two orbiting propellant stations. The seventy-two orbital hotels will include ten that are in polar orbit (which, over the course of successive orbits, will allow guests to view the entire world) and two in elliptical orbit, giving guests both close-up and distant views of Earth. He anticipates daily scheduled flights to the Moon, where there will be hotels and resorts at the lunar poles (where water is available). He expects that at that time there will be 5 million passengers per year, and at any one time seventy thousand people will be in orbit.

FITTING IN

If the visions of the space tourism advocates are realized, then space will become home to hordes of visitors each year. If and when tourists visit places such as the Moon, we may expect the kinds of problems that occur when large numbers of tourists show up at formerly isolated locations. Tourists do not always merge well with the people who are already there, and either due to ignorance or callousness they pose some level of environmental risk.

Tourists and Professionals

Even given the space tourism industry's optimistic time line, space tourists could be beaten to the Moon by explorers, scientists, and miners. Based on experience in Antarctica, we can guess how the professionals and the tourists will get along. Until relatively recently, there were only three ways that a North American could get to Antarctica: as a scientist supported by the National Science Foundation, through a navy assignment, or by employment with one of the civilian contractors that provided maintenance services at various Antarctic posts. But this has changed.

At first, cruise ships brought tourists to the coast of Antarctica. Now, cruise ships and special flights take tourists to some of the coastal stations and even the South Pole itself. From the professionals' point of view, tourists impose on their valuable time. Scientists may feel obligated to clean up or police the area before the tourists arrive, and to watch out for the tourists and answer their questions. Despite careful instructions to the contrary, even well-meaning tourists can disrupt an experiment. Then, too, the professionals feel that they have earned the right to be there, while the tourists have merely spent money to do so. Finally, despite the huge mass of the Antarctic continent, tourists can make the place too crowded—at least when they all try to jam into one hut. Tourists, in other words, can impinge on the professionals' privacy and time and chip away at their sense of distinctiveness and accomplishment.

There is, of course, a relatively simple solution. Although the industry will require professionals to pilot the craft and maintain the first space hotels, the tourists could be confined to special locations. The spacefarers that tourists come into contact with will not be scientists but will work for the tourist industry. If they can be sequestered in a specific tourist haven, tourists will not interrupt scientific experiments or keep space industrialization from continuing apace. This separation may not be so easy when, for example, tourists visit the Moon or Mars and want to examine an industrial site or visit a historic area, such as the sites of the Apollo landings.

Certain areas may become protected monuments. The footprints and debris of the first astronauts on the Moon might be protected for all time, as, no doubt, would be the remnants of the first lunar and Martian colonies.

Environmental Protection

Tourism increases the number of people in an area, and this alone can alter the local ecology. Tourists who are well intentioned inadvertently kill things and leave trash behind. Tourists who are not well intentioned snatch botanical or geological specimens for souvenirs, leave trash where it falls, and otherwise hurt the environment. This has occurred in Antarctica, and it could very well occur in space.

Here, we may expect to learn from the experiences of the National Park Service. Essentially, there are two conflicting strategies—development and protection—that in combination satisfy most people's interests. Some areas are developed or even "enhanced" to provide safe, comfortable, and interesting areas for the general public. These are the highly accessible visitors' centers that are usually located near special attractions and offer great vistas. Such areas have been engineered to be safe and comfortable, with paved pathways or plank boardwalks, handrails, illuminated parking areas, and the like. They are continually staffed during operating hours and have such amenities as displays, running water, and toilets. Other areas, wilderness areas, are highly protected and left alone to be modified only by the processes of nature. If visitors are allowed at all, they are expected to have certain levels of knowledge and proficiency, are subjected to appropriate rules and regulations, and may be led by professional guides.

CONCLUSION

Advocates of space tourism paint a very appealing picture. Much of their attention is devoted to developing an economical, fully reusable launch vehicle. If in fact they can reduce launch costs to a small fraction

of what they are today, they will eliminate one of the greatest bottle-
necks to our advancement in space. Not only would this open the door
to tourists, it would provide the cheap transportation needed to send
workers to construct solar power satellites, mine the Moon, and reach
many of the other goals for commerce in space.

By raising the prospects of space tourism, industry advocates give us
a wonderful gift. People are intrigued by the idea of spaceflight, but their
excitement is restrained because they don't consider it a possibility for
themselves. Not only would making space accessible to a broad segment
of the population give people exciting new experiences, it would encourage many different kinds of human activities in space. Thus, the space
tourism industry could develop both the technology and the popular
support required to accelerate human progress in getting off our planet.

As is so often the case when we consider our possible future in space,
the visions seem to rest upon wishfulness and optimism as well as any
economic and engineering data. Part of the problem may be that we
are so conditioned by the past forty years of space exploration that it
is almost impossible to conceive of space travel as inexpensive and
popular. There seems to be a fundamental disconnect between the tens,
perhaps hundreds, of millions of dollars that it costs to put a person
into space today and the ten thousand dollars we hope it will cost to
put a person into space at midcentury.

Even if it is true that concerted efforts mounted today could bring
spectacular results in a decade or two, very few companies are trying
to mount such efforts. Only recently has NASA viewed space tourism
as a viable possibility, and the aerospace giants seem content with
launching satellites and fulfilling government contracts for largely expendable equipment. Their perspective may broaden, however, as defense contracts dwindle and as our skies become saturated with communication satellites, leaving these companies with tremendous
capacity to turn to new ventures.

Space camps, theme parks, and virtual experiences are already available, and suborbital flights are on the horizon. It is possible that we
have a basically accurate outline of the future of space tourism beyond
that, but that the time line is optimistic. Those of us who would like
to visit space should wish the industry Godspeed. We live in a fast-
paced society and are accustomed to very rapid technological change.
Nonetheless, it took many centuries for the dugout canoe to evolve
into *The Love Boat*.

CHAPTER 13

SPACE SETTLEMENTS

Mercury, Gemini, Vostok, Apollo, Skylab, and Salyut are behind us. We are almost two decades into the space shuttle era, and the ISS is in orbit. We know how to return to the Moon and take our first steps on Mars; these ventures remain in the future but seem tantalizingly close. Let us peer deeper into the recesses of the crystal ball, to those distant times when people may be living on the Moon and Mars and when huge orbiting colonies may house more people than live on Earth.

As in the case of space tourism, the discussion of space settlements and (in the next chapter) interstellar travel is necessarily somewhat speculative. Although many of the proposals that I will review here include engineering specifications and detailed explanations of the technology, my focus remains on the human element. From my perspective, if space settlements materialize as they are currently envisioned, they will be less interesting as engineering triumphs than as human accomplishments that will shape the lives of future generations. Space settlements are intended to solve human problems. If they evolve as we hope, they will offer safe, provident, and wholesome physical environments; political and social reforms; and abundant opportunities for residents to flourish materially and psychologically. A strong humanitarian bias contributes to our vision of space settlements.

VISIONS OF THE FUTURE

Until the close of the nineteenth century, the Western frontier absorbed population growth and offered people hope and renewal. Today, space beckons as the new frontier. If only we could tap them, space has the resources to give billions of people, including those accustomed to poverty, a comfortable life. Space hospitals and retirement homes could offer older people sanctuaries where microgravity reduces the burdens on their bodies and allows them to discard walkers and canes. Space prisons could isolate criminals whom we cannot allow to escape. Settlements may be great locations for think tanks that require the highest degree of secrecy and impenetrability from the outside.

Space could be useful for industrial processes that would create deadly environmental conditions on Earth. It is an ideal location for high-risk experiments and industrial processes that nobody wants in their own backyard. Moreover, space settlements would provide a first line of defense against incoming meteors or missile attack. They also would offer humans an opportunity to continue as a species as Earth runs out of resources, after a large-scale natural disaster such as a meteor strike, and after the Sun dies. And, if you believe in extraterrestrial intelligence and the feasibility of interstellar travel, an orbiting satellite or planetary outpost would be a better setting for initial "face-to-face" contact than any place on Earth.

On Earth, nations participate in a zero-sum game, where any territory won by one nation is lost by another. In space, through such means as disassembling a large asteroid and transforming it into a large orbiting community we may be able to *create* new land. On Earth, nations nudge up against each other. In space, colonies can be dispersed so that like-minded individuals can band together to experiment with different political systems and lifestyles without upsetting their neighbors. Unlimited resources and territory coupled with spatial buffering from ideological conflicts would reduce some of the traditional bases for war.

What will the first settlers be like? Traditionally, military personnel have been the first to enter new, unusual, and potentially dangerous environments. In recent times, scientists and entrepreneurs have come next. We might expect that when emigration begins in earnest, space will draw people who are restless and who see opportunity for a better life. The same sort of people who stow away on ships or scale walls to find freedom and economic security on Earth may be among the first wholesale volunteers for space. We might also expect adventurous people who want to do something worthwhile: the same kind of people who enlist in the Civilian Conservation Corps or Peace Corps. But in the long run, space is for everyone.

Moonbase

Lunar colonies are championed by space industrialists, who see many resources there and find strong economic justification for a permanent return to the Moon.[1] They are also advocated by NASA scientists, who argue that a return to the Moon is a logical next step in the human occupation of space.[2] These scientists suggest that, although we have the technology to go to Mars, we need greater experience on the Moon to prepare ourselves better for the Mars trip. A rejuvenated lunar program would help us develop spacecraft and other systems that will remain trouble-free for the three years required for a Mars mission. It would also allow us to increase our understanding of the biological and psychological risks of crewed Mars missions and develop more effective countermeasures to physiological deconditioning, psychological stress, and other threats. A return to the Moon could constitute a rapid success that would bolster overall support for space exploration and actually accelerate our progress to Mars.

Although NASA has no plans for a return to the Moon, the Artemis Society International has mounted the Artemis Project to establish a permanent, self-supporting community on the Moon. In Greek mythology, Artemis is the moon goddess, the twin sister of Apollo, who stands for the sun. Privately funded through investments in established industries such as movies, television, publications, and merchandising, the society intends to show that spaceflight is within the reach of private industry, develop lunar resources for profit, and encourage commercial activity in space. Their lunar community would support lunar exploration and scientific investigation, lunar industries, homesteading, and tourism. This would begin with rugged expeditions for the kind of people who like mountain climbing, spelunking, and safaris, and eventually will grow into luxury-class trips on giant spaceliners. The project is based on the shuttle and other known technology, and the society hopes to establish its first camp on the Moon within the decade.

What would a full-blown community on the Moon be like? For over forty years, scientists, engineers, and space planners have developed plans—sometimes very comprehensive and detailed plans—for bases on the Moon.[3] Some of this thinking is reflected in Ben Bova's imaginative book *Welcome to Moonbase,* which gives us a useful reference point for discussing life on the Moon.[4]

Bova anticipates an initial return to the Moon in the opening years of the twenty-first century, followed by the evolution of full-fledged settlements over the next few decades. These communities would not appear full-blown overnight. Whether we are looking at a rough work-camp, a community of a few dozen people, a city of a few hundred or thousand people, or a fully inhabited world housing millions depends on our time frame.

Bova picks up the story in the mid-twenty-first century, twenty or thirty years after humans return. At that time perhaps five thousand people could be on the Moon, at Moonbase and its counterpart, Lunagrad, and at tourist attractions scattered here and there. Moonbase would house about two thousand people, who would be occupied predominately with mining and manufacturing and would work under direct control from Earth.

Moonbase would be constructed of large, underground shelters protected from damaging radiation by thick covers of regolith. A huge plaza (600 meters long and 75 meters high) and clusters of other buildings would contain private living areas, common gathering and recreational areas, restaurants, theaters, and shops. Although there would be no windows in these subterranean dwellings, outside cameras connected to giant video screens would allow occupants to peer at the landscape, at least when the screens were not in use for watching TV or playing video games. Interiors would have very high ceilings and be arranged in other ways to convey as much of a sense of spaciousness as possible. To further enhance habitability, designers would "green" the quarters with plenty of plants.

Administered by a central authority, settlers would come from many nations and retain their national citizenship. A multinational corporate authority would control operations and have the right to maintain the peace and enforce safety regulations. The inhabitants would get to vote on matters regarding community life. In a dramatic departure from other visions, Bova's settlers would pay income taxes to their home countries.

People who chose Moonbase would work there on one- or two-year renewable contracts. After a period of five years, they could apply for permanent residence. Of the two thousand colonists, approximately 10 percent would be permanent residents. Workers on short-term contracts would come alone, but those who had been accepted for extended employment would be allowed to bring their families along.

Bova sees Moonbase as being organized into three departments: management, health and safety, and technical services. It is the last of these that would serve the basic productive functions, including mining and manufacturing, space transportation, tourism, exploration, and research. An appealing aspect of Bova's vision is his inclusion of opportunities not only for the usual scientists and engineers but also specialists in business, health and human services, tourism, and the arts.

Mars

For over half a century scientists have pondered how to get us to the red planet. According to Robert Zubrin and Richard Wagner, most of their scenarios call for enormous expenditures over many years.[5] They require us to wait until we can construct large, posh spaceships to get us there in style or speedy propulsion systems to get there in a very short time.

Under most scenarios, part of the immense cost would be building construction facilities in low Earth orbit to assemble the spaceship. Another part of the expense would be maintaining a "mother ship" in Martian orbit while a small team descends to the surface of the planet, just as Apollo teams once descended to the Moon. Under most scenarios we would not get much scientific return on our investment because the crew would stay on Mars for just a few months. This would be done so that Earth and Mars would be in favorable positions relative to one another for the trip there and back.

Zubrin's plan, called "Mars Direct," proposes a piloted Mars mission requiring about ten years from approval to departure and costing about the same price as introducing a new major weapons system. Mars Direct rests on the idea that we can build launch vehicles of sufficient size that we can hurl them into orbit, send them on to Mars, place them in Martian orbit, and then land them. It would not be necessary to build construction facilities in Earth orbit nor to keep a mother ship in Martian orbit. The entire crew would participate in surface exploration.

Mars Direct would use small, lightweight spacecraft. Prior to the crew's departure from Earth an automated ship would arrive at Mars and use raw materials to manufacture fuel for the return trip. When it came time for the spacefarers to return home, after about five hundred days on the surface, they would simply hop into the fully fueled ship and take off. Mars Direct proposes a series of interlocking missions, alternating between robot and human flights.

The first crews would consist of four people. This includes two mechanics, who are good at diagnosing and fixing the technical systems, and two scientists, who have expertise in such areas as geology and exobiology. By limiting the crew to four, each piloted flight would cost much less than those that call for crews of seven or eight.

Over time, many settlers would arrive on Mars. At first, Mars Direct settlers would live in the spacecraft that brought them there. Initially, these large "tuna fish cans" would be spread out across Mars, so that different teams could explore different parts of the planet. Later they would be clustered together and connected by tunnels to form the nucleus of the colony. Inflatable structures for working areas and agricultural production would be added. Permanent buildings constructed of Martian bricks would appear. Ironically, because of this brick architecture, the first permanent buildings on Mars may be reminiscent of early Rome. If all goes well, over the centuries terraforming would raise the temperature, liberate water, and thicken the atmosphere, perhaps transforming Mars into home for millions of people.[6]

Mars Direct is attractive because it offers a fresh approach for reaching the red planet. It is creative and economical and it suggests realistic ways to negotiate many of the obstacles that keep us on Earth. Perhaps most significant, it could be accomplished within a time line short enough that most people alive today would be able to witness the landing of the first humans on Mars.

Orbiting Colonies

In the 1970s, Gerard K. O'Neill wrote that the Moon and solar planets are not necessarily the best destinations for wholesale emigrations from Earth.[7] It is expensive and time consuming to get there, especially to Mars. Sunlight, the source of power and life, will not be readily available during the two-week lunar night, and it will be difficult to collect on Mars and other planets that are farther from the Sun than Earth. Terraforming, if it works at all, will take centuries to produce good results. The overriding problem is that planets and moons can hold only a small fraction of human population growth. To accommodate such crowds, O'Neill proposed huge orbiting communities located between Earth and the Moon.

German rocket scientists foresaw large space stations as fueling stops for rockets and as places for constructing vehicles for trips to the Moon and beyond. In 1929, Hermann Noording developed the idea of a large wheel-shaped satellite—reminiscent of the space station in the movie *2001: A Space Odyssey*—and in the 1950s, Wernher von Braun developed a similar plan.[8] But it was O'Neill, a Princeton physicist with a strong social conscience, who saw huge orbiting satellites as salvation for Earth. Overcoming the "giggle factor" that plagues scientists who try to enter uncharted waters, he garnered NASA support, organized a series of breakthrough workshops, and set forth a detailed rationale for space colonies in his book, *The High Frontier*.[9]

Like most other proponents of large-scale emigration to space, O'Neill believed that the world, with its burgeoning population growth, was entering an era of decline. He noted the rapid depletion of fossil fuels and other resources and growing concerns about environmental pollution and global warming. Like Krafft Ehricke before him, he rejected "think small" solutions—that is, limiting population growth or restricting the use of Earth's resources.[10] These would lead to technological and economic stagnation and (because people would be afraid to risk remaining resources by trying something new) the suppression of new ideas. The climate of scarcity could raise frustration, perhaps triggering riots and war.

O'Neill set forth detailed, phased plans for developing orbiting colonies. The construction team would launch only a few of their raw materials from Earth. Most would be mined on the Moon. There, perhaps two hundred spacefarers would develop simple, highly automated strip-mining operations. Using a device known as a mass driver, lunar material sliced into shapes reminiscent of large, thick poker chips would be accelerated electromagnetically along a long track, break free of the Moon's weak gravity, and fly through space to be caught at the construction site. There they could be used like bricks or transformed into other types of building materials.

O'Neill envisioned building three colonies or "islands," each of which would be rotated or spun to create artificial gravity. Island I would be a sphere about 1 kilometer in diameter and would house approximately 10,000 people. Island II, a relatively short cylinder 1.8 km long with rounded ends (sometimes likened to a cold pill) would house 140,000 people. Island III, a cylinder 3.4 kilometers in diameter and 52 kilometers long, would have 400 square kilometers of living area and house 10,000,000 people. Eventually, large orbiting settlements might consist of two cylinders, one for living quarters and one for work. In this way the undesirable risks and by-products of research and manufacturing would not affect the cylinder where people dwelled.

O'Neill's contributions to the development of space settlements are more than explorations of the physics and engineering involved. He moved space colony design into the realm of the possible and attracted the professional interest of architects, scientists, and engineers. This interest was sustained in later NASA Ames projects that led to different designs, such as a large "paddlewheel" satellite. Furthermore, O'Neill's focus on population explosion, environmental decline, and other problems that gained salience during his era, coupled with his engaging, informative, and lavishly illustrated book, attracted the support of many people who had never before given space settlement serious thought.

The Millennial Project

In his 1994 book *The Millennial Project*, Marshall Savage offers a daring proposal for colonizing space in "eight easy steps."[11] The appeal of his work lies in part in his highly crafted, detailed analyses and arguments, and in part in awesome prose that evokes images from subjects ranging from the Norse Vikings and King Arthur's Court to cutting-edge science and technology. He has a bold vision of how we can assemble the resources that we need in order to occupy space. It rests on the twin pillars of raising capital and finding novel ways to simplify technology and lessen costs.

The first step is developing a prototypical settlement on Earth. Known as Aquarius, this artificial floating island would be located in international waters, where residents would be free of repressive laws and taxation. Most of the structure would be fabricated from natural substances extracted through electrolysis and accreted to form "seament." Power would come from ocean thermal energy converters (OTEC) that generate electricity by mixing abundant quantities of frigid water, pumped from far below the surface, with giant quantities of warmer water pumped from above. Algae and other forms of aquaculture (including delicacies such as shellfish and lobsters) would feed the colonists. Exported electric power and sea products, including food and cultured pearls, would raise money for the next steps toward space.

Profits from Aquarius would be invested in a reliable, inexpensive launching system that, in contrast to present-day rockets, would not consume tons of useful chemicals and harm the environment. Savage proposes a long, horizontal underground tunnel that eventually turns upward and exits at a mountaintop. Using OTEC power, launch vehicles would be accelerated electromagnetically along the horizontal track and would build up tremendous acceleration before they emerge. At that instant, ground-based lasers also powered by the OTECs would fire at the rear of the vehicle, where they would vaporize a thick layer of frozen material, pushing the craft to orbital velocity.

The first pioneers would live in a crude construction shack orbiting Earth and venture forth to build the first settlement, called Asgard. There, habitats and enclosures would consist of transparent membranes arranged in spherical forms. Asgard would be a staging ground for later settlements on the Moon and Mars. Enclosed in giant bubbles and situated in the large craters that constitute the valleys of the Moon, lunar (and later Martian) communities could grow into enormous sizes. Characterized by diversity of structures and lifestyles, they would be connected by efficient, high-speed transportation systems.

For Mars, Savage stresses the need for vigorous terraforming that will yield relatively speedy results. These include importing water by crashing fragmented icy comets on the planet's surface. Over time, he hopes, Mars will become the site of golden, domed cities surrounded by luxurious grassy fields and knolls. Later, settlers would fan out into the asteroid belt and set forth for neighboring stars.

The Millennial Project takes us, then, from a small ocean-based colony to interstellar migration. Projections call for an autonomous floating city somewhere around 2012, and interstellar migration between 2500 and 3000. When we review this and other proposals, we should not become too obsessed with details, for they are sure to change. Savage's contributions include articulating an integrated plan for the step-by-step occupation of space; forcing us to attend to ways to fund our movement into space; and urging us to develop simple, economical, and in some cases, radical new technologies.

LIFE ON THE HIGH FRONTIER

Many proposals for space colonies are beautifully written and lavishly illustrated. Based on a combination of science and art, they are mixtures of reality and wish. They offer the lure of a better life, including pleasant surroundings, improved political forms that thoughtfully balance the needs of the individual and community, supportive social relations that rest on personal trust and caring, tremendous individual freedom, and increased opportunity to find happiness and material wealth.

Existence Needs

Existence needs are satisfied by safe, comfortable, and provident environments that accommodate people's biological requirements, engender a sense of security, and offer material gain. Although some building materials must be exported from Earth, settlements would be constructed from metals and regolith mined on and scraped from the Moon and Mars and, later, from disassembled asteroids.

Bova and O'Neill propose shielding settlers from radiation by means of lunar rocks or cast by-products of lunar refining. Savage envisions shielding orbiting communities with massive amounts of water sandwiched between huge, spherical or hemispherical transparent membranes. On Mars, shielding requirements will be less onerous, because even the thin Martian atmosphere affords some protection against radiation, but here again regolith will be put to use.

Space junk and meteorites threaten orbiting and lunar communities, but because of the habitat's enormous volume even a relatively large object striking it would not lead to an explosive loss of atmosphere. At Moonbase, Asgard, or on Mars, the loss of atmospheric pressure would be so slow that, in the absence of a network of sensitive sensors, occupants would barely notice the drop for several days. There would be plenty of time for repairs.

O'Neill proposes rotating or spinning his islands to create artificial gravity in order to protect the residents from deconditioning. Savage would not use artificial gravity, but hopes to keep Asgard colonists healthy through electrical stimulation of the muscles. The Moon and Mars offer some gravity, but much less than Earth's. Without special regimens it might be very difficult for long-term residents of the Moon to visit Earth. People born on the Moon might not be able to visit Earth except under unusual conditions.

Orbiting settlements would be free of terrestrial geological disasters such as earthquakes and inclement weather, including storms, monsoons, droughts, heat waves, and cold snaps. Hopefully, we could prevent the importation of undesirable insects and other vermin from Earth. Clean technologies could let us avoid pollution and minimize problems associated with environmental health. Because there is no atmosphere, the weather on the Moon would be predictable, but settlers would have to put up with dust storms on Mars.

Crowded and primitive at first, living quarters would become increasingly spacious and pleasant as the settlement matured. Savage foresees efficiency apartments and family units that are larger than many residences on Earth, although colonists might have to choose between spacious quarters in the interior of the settlement and smaller peripheral apartments that offer spectacular views. O'Neill's plans avoid high-rise structures and call for comfortable, stacked, low-rise buildings with plenty of shared or common areas. He envisions lush vegetation, lakes and rivers, small villages, and intermediate-sized cities (approximately the size of San Francisco) on Island III.

Food should be abundant, healthy, and tasty. Tomorrow's space-farers may produce such novelties as yeast burgers on the Moon and mushroom burgers on Mars. As for staples, Savage expects production of algae-based protein made palatable by varying textures and adding synthetic flavors. Drawing on then-recent developments in production agriculture, O'Neill suggests that, using mirrors and shutters to bathe the crops in continuous sunlight, little square footage would be re-quired to grow food on a per capita basis. Because it takes tremendous acreage to raise cattle, beef would be a rarity, but in addition to healthy vegetables settlers might produce chicken and turkey and perhaps pork.

Space settlers would have jobs and the opportunity to make money, perhaps more than they could make on Earth. One stream of income, foreseen by both O'Neill and Savage, would result from information processing, consulting, software development, and other highly com-pensated activities that require very little office space and that can be done by telecommuting. Unfortunately, because it is a lot more expen-sive to maintain a computer programmer in space than on Earth, this kind of employment would be at best a stopgap measure.

Another income stream would derive from the goods and services that settlers would provide one another in space. Settlements would have their doctors, lawyers, maintenance engineers, entertainers, and fast-food merchants. Of course, none of these activities would be lu-crative until large numbers of people are actually there.

Another income stream would come from exports. As noted in chapter 1, exports could include solar electric power beamed to earth, helium-3 for use in fission reactors, and minerals mined from planets and asteroids. Zubrin describes an economic triangle involving Earth, Mars, and the asteroid belt: Earth manufactures and exports high tech-nology to Mars and the asteroid belt. Mars exports basic supplies and low technology items to the asteroid belt, and the asteroid belt returns metal and other resources to Earth.

Settlers would be attracted to Bova's Moonbase by the opportunity to earn high salaries (set at 25 percent above that for comparable work on Earth) and excellent working conditions. Settlers might stay on the Moon or Mars for an extended period of time since it makes more sense to grant high salaries than to pay for a continuous shuffle of workers back and forth. The Moon, like the Arctic before it, could become an excellent place for adventurous, talented, and work-oriented people to stash away large amounts of money for advanced education or for starting a business following their return to Earth.

Most of the work at Moonbase would be done indoors in carefully controlled, comfortable, pollution-free shirtsleeve environments. Robots and automated systems would do the mind-numbing, repetitive, and dirty tasks, and most of the work done outside. All of this is suggestive of industrial humanism of the 1930s. The employer pursues a policy of benevolent self-interest, using high pay and good working conditions to attract the most capable and most motivated workers to do the job.

Relatedness Needs

Relatedness needs are satisfied by warm, caring interpersonal relations; organizations that treat employees and customers or clients as unique, worthwhile individuals; and benevolent, nonoppressive governments that take their citizens' welfare to heart. Most of the literature on space colonies focuses on relatively small communities (involving anywhere from a dozen to a couple hundred settlers) or very large communities, with hundreds of thousands or millions of residents. The challenge for small settlements lies in developing governmental and other institutions in a setting that is basically understaffed. The challenge for large settlements is in ensuring that these do not evolve into totally impersonal entities where citizens get lost in the crowd.

Richard Terra imagines social life in small groups of asteroid miners in the Oort Cloud at the outer fringe of our solar system.[12] The Oort Cloud consists of tens to hundreds of billions (if not trillions) of comet nuclei ranging from .10 to 10 kilometers in diameter and each weighing between ten million and ten trillion metric tons. It surrounds us like a hollow shell whose inner and outer limits are approximately 1 to 2 light-years from here. The nuclei contain frozen gasses, organic chemicals, and metals. The prospectors' biggest technical problem will be finding a suitable energy source (perhaps nuclear fusion, perhaps highly focused starlight) that far from our Sun.

Terra foresees initial groups of one hundred to two hundred settlers forming colonies in various sectors of the Oort Cloud. These settlers would have to be extremely rugged and self-reliant. They would be forced to pool their resources and efforts to survive and prosper. Of necessity, they would be highly cooperative with one another. Everyone would have considerable latitude for individual expression (so that they can find happiness), but it is very unlikely that anyone would become so individualistic as to seriously antagonize someone else. Such small communities would be unlikely to become highly stratified. More likely they would be run like families. Important decisions would be reached by consensus, and Terra thinks that the community would avoid choices that create special hardship for specific individuals. There would be a strong sense of common fate, and there would perhaps be some elitism when the settlers compared themselves to people in larger settlements.[13] Precedence for this is found in the Antarctic, where some residents of small, isolated stations consider themselves superior to the hordes at McMurdo and other large staging bases.

About the same time that O'Neill was planning huge satellites, the world was becoming aware of serious problems associated with crowding in our cities. Solid, middle-class citizens were discovering themselves stalled in freeway traffic and jostled on city streets. Some people felt like they were faceless entities cast adrift with the anonymous hordes. Sociologists describe this as *anomie*, a state where people feel disconnected from one another and alienated from society as a whole.

Experimental research of that era underscored some of the detrimental effects of crowding. J. B. Calhoun compared crowded and uncrowded rats.[14] Those pressed together in close confines had difficulties courting and suffered from sexual dysfunction. They produced fewer offspring, and these offspring were puny and sickly. Normally, rats are careful nest builders. Crowded rats were less likely to build nests, and, when they did so, their nests tended to be incomplete or of inferior quality. The crowded animals were highly aggressive and prone to illness, malfunction of their organs, and tumor growth, and they had higher mortality rates. Calhoun used the term "behavioral sink" to describe the crowded pens that undermined health and caused maladaptive behavior. Studies by Calhoun's contemporaries hinted that high population densities had adverse effects on humans, including poor work, physical and mental health problems, and increased crime and delinquency.[15] The problem before us, then, is how to bring the expected benefits of the smaller settlements to giant settlements that are home to millions of people.

One tactic is to use design techniques that increase a sense of spaciousness and break up the interior of a large settlement, so that it is transformed into a collection of smaller communities. Thus some plans call for distant horizons; visible, interspersed structures; and the use of colors and lights to open up areas. There may not be as much room for grassy fields as we would like, but there is ample allowance for vegetation, including trees, shrubs, and hanging plants. To create a friendly look, buildings can be aligned at odd angles, rather than with military precision.

Architecture can foster positive relationships among settlers. The clustering of buildings, orientation of entrances and exits in different ways, and development of neighborhood parks and other common areas are intended to make it easy to meet, mingle, and develop a sense of community. Variations in architectural styles reinforce a sense of community within a particular subdivision and give residents of different subdivisions their own sense of identity. The models for some space settlements are quaint sections of European cities whose funkiness makes them pleasant despite high population density, and the models for others are the small, carefully planned, ecologically minded and occupant-friendly housing developments that began to dot the North American landscape in the 1960s and 1970s.

Even as visionaries sought to avoid crowding, they tried to escape the blight of big governments. Governments should be kept to a minimum and should interfere in individual freedom as little as possible. Each person should have the right to do as he or she pleases, with the proviso that these activities cannot infringe on other people's rights. The goal is to establish minimalist, low-profile governments that meddle as little as possible, and, when they do meddle, they should try to do so in understanding, friendly ways.

The space settlement literature contains few references to monarchies, dictatorships, and other authoritarian forms of government, but many references to democracy, so the settlers themselves will be in charge. (An exception is Bova, who sees residents of Moonbase beholden to authorities on Earth.) O'Neill claims that he is "a-political" (while arguing for liberal and humanistic forms of government), whereas Savage urges a pure democracy. A representative democracy, notes Savage, tends all too often to be perverted to accommodate the interests and needs of the elected officials and their cronies. On an advanced Millennial colony, decisions will be made by millions of individual minds working in concert, achieving synchrony, and, in the process, perhaps approaching a higher level of consciousness. Everyone who is able and chooses to do so should vote on each issue, which might be done electronically in a form of government that Jim Dator calls "teledemocracy."[16]

With the possible exception of politicians and officious government bureaucrats, a minimalist democracy offers something for everyone. The poor and dispossessed of the world become enfranchised and gain some control over their lives. Middle-class people can live without having to support the indigent and without mistreatment at the hands of large governmental bureaucracies. Wealthy people can see increased freedom to pursue entrepreneurial goals, unfettered by needless laws and regulatory agencies. If the vision actually materializes, then almost everyone will win.

We hope that space will draw decent, hard-working, honest people, and that, thanks to prosperity and a benevolent government, crime will be left behind on Earth. But this is unlikely to happen. Alvin Rudoff points out that space settlements will reflect their inhabitants' social backgrounds, including their weaknesses as well as strengths.[17] Misbehavior, he notes, could include arson, gambling, substance abuse, pornography, suicide—all manner of crime and delinquency, including sabotage and terrorism. The designation of alcohol, drugs, firearms, or other items as contraband could result in an "underlife"; that is, organized illegal behavior kept secret from authorities or those who seem likely to disclose the behavior to authorities. In space, as on Earth, we can expect laws and enforcement procedures.

Although particular laws may strike us as silly or wrong, put to the test most people support law and enforcement procedures that prevent "war of all against all" and protect societies and individuals from excessive harm. Space settlements, at least those not micromanaged from Earth, will allow a fresh start. As we move to new worlds we can reassess our laws. If we so choose, we can discard laws that relate to victimless crimes, perpetuate discrimination or inequity, or impose punitive taxes that prevent economic growth. We may want to preserve the right to live unharmed by others, to protect oneself, to express one's views freely, to make choices, and to receive restitution when so justified. The underlying values are individuality and freedom tempered by the need to protect the greater good.

Legal systems proposed for space tend to take one of two directions. The first, based on a fundamentally optimistic view of human nature, focuses on mediation and arbitration. The model here, according to Donald Scott, is the community arbitration panel that works to restore amicable relations among quarreling neighbors.[18] That is, the colonists could establish a panel to ensure that people understand the consequences of their acts, make restitution when possible, and learn to get along with one another.

The second model, based on a less optimistic view of human nature, is military justice. Kenneth Schwetje points out that whereas a court martial preserves a defendant's basic rights and follows certain civilian procedures, it follows simplified procedures.[19] Military justice does not depend upon a huge legal apparatus that allows appeals at multiple levels, and could be easily adapted to space.

Dator theorizes that the primary concern will be the peace of the community, with relatively little weight assigned to the "rights of the accused," "the rights of the victim," and "restitution."[20] Small colonies may also be shorthanded, and so neither incarceration nor the death penalty would be attractive. The most obvious alternatives are reduced privileges, heavy fines, and deportation back to Earth. Someday there may be penal colonies, including a few that, like Australia, evolve into highly successful independent societies.

If they meet expectations, then space settlements will reduce social inequities, restore individual freedom, and guarantee the right to pursue happiness. Although they may consist of huge numbers of people, they will avoid crowding by spreading the population around a number of smaller communities. They also will eliminate the unpopular aspects of big government. They will be intended to keep people from feeling disconnected with one another and alienated from society as a whole. Here, we have an interesting combination of a reversion to small-town rural America but with mammoth populations and high technology.

Growth Needs

Space settlements are intended to help people to feel secure and worthwhile, and to grow psychologically. Freedom from want and a supportive social environment provide good starting points. Robert Zubrin, especially, stresses that the first generation of emigrants to a new colony will see themselves as having a "fresh start" in a location where they will be judged by their merits, not by such factors as their background, gender, or ethnicity.[21] Drawing on the work of historian Frederick Jackson Turner, who sought to understand the social consequences of westward expansion during the nineteenth century,

Zubrin draws parallels between opening the western frontier in the 1800s and opening the next frontier in space. According to Zubrin, our Western humanist society, with the premium that it places on the individual and a high rate of technological change, is a direct outgrowth of the opening of the western frontier. Today, however, the western frontier no longer exists. Wealth is concentrated in the hands of a few, and we are no longer in a period of rapid technological growth.

Opening up the high frontier, states Zubrin, will create new wealth. On Earth almost everything of consequence is owned, and a "have" must surrender something if a "have not" is to gain. By making new resources available, the frontier creates new opportunities. Pioneers in space can stake claims to land, homestead, and create new wealth out of the resources found in space.

Given few hands and the challenges of a frontier, people are forced to find new ways of doing things, to develop new laws and customs as well as new technologies. The labor shortage reduces status distinctions. On a frontier, the physician, electrician, and carpenter's apprentice are all worthwhile. The premium placed on the individual generates a greater willingness to invest in education and welfare. This, in turn, unleashes creativity and talent.

When people move to a frontier they leave old institutions behind. By opening up the space frontier we will foster self-reliance, education, and equality, and thereby set the stage for dynamic democratic forms of government. Because space offers ample room for future expansion, we could have endless renewal on the high frontier.

Not everyone agrees with this attractive analysis. Some historians are rethinking Turner's original work. Howard E. McCurdy points out that the frontier did not necessarily provide equal opportunity for all—blacks and Mexicans undoubtedly were short-changed, and it is unlikely that many Native Americans would share Turner's opinions.[22] Moreover, some of the benefits attributed to life on the frontier may have been due to other factors, such as the personalities of the people who chose to emigrate. Nonetheless, space opens up new opportunities, and let us hope that settlers will take good advantage of their fresh starts.

Space settlements can accommodate human diversity and preserve cultural differences, as well as encourage the development of new space-based cultures. Since space is vast, there can be many settlements, each tuned to special tastes. To be sure, there are subcultures within the nations of Earth, but these are to a large extent constrained by the dominant culture. Whether we are talking about architecture, political systems, or lifestyles, large orbiting communities (or perhaps even entire planets) could be devoted to a particular way of life.

One of the strongest proponents of variety is Magorah Maruyama, who proposes architectural designs that will express the colonists' values and accommodate their preferences and behaviors.[23] For example, one such settlement might feature completely separate dwelling units, each essentially sealed off from the others and allowing a very individualized and private life. Another settlement might be designed in a much more open fashion, with communal living areas that promote group activities. In the former, people can enjoy hard architecture that serves to isolate and defend them against unwanted intrusions, while in the latter residents can enjoy softly flowing architecture that promotes a sense of openness and accessibility.

CONCLUSION

Proposed space settlements have attempted to rectify the problems salient at the time they were planned. These problems included a growing concern with the environment and with big government, and a sense that social relationships were becoming too impersonal. As envisioned by some of their planners, space settlements offer hope for repairing and revitalizing human society and curing anomie, ennui, and even (thanks to artificial textures and synthetic flavors) lunch pail lassitude! Space, in other words, provides the opportunity to conduct social experiments and construct an ideal society. Not everyone agrees with this. A thoughtful and dissenting note comes from sociologist Alvin Rudoff.[24] In his view, when we look forward to space settlements, the search for Utopia will be no more fruitful than it has been on Earth. Instead of deliberately experimenting, we should establish an environment that permits natural processes to evolve. We should not seek to establish a "static perfection" but rather a dynamic society geared to adaptability, adjustment, and change.

In their 1986 book, James and Alcestis Oberg include a NASA artist's conception of a space station and an actual interior view of a Salyut station.[25] The flowing lines, spaciousness, freedom from clutter and overall aesthetic appeal of the imaginary space settlement represent perfection when compared to the cramped, functional, cluttered look of the real thing. It may be possible to construct large, attractive habitats, and perhaps tomorrow's space settlements will be a far cry from contemporary space stations. Nonetheless, we should expect many slips between design and execution. Furthermore, mass emigration is at least decades, maybe centuries, away (if it occurs at all). Between today's planning efforts and tomorrow's wholesale departures from Earth, new scientific discoveries and technologies will emerge. As we become more knowledgeable and capable we will change our plans. At one time we believed that the Moon was covered with a thick layer of dust. A Moon base planned in that era was designed to float in this stationary ocean of dust, held in place by cables and anchors.[26] What else will we have to rethink, and will our job become easier or more difficult?

Perhaps in our efforts to envision tomorrow's space settlements we can be compared to a naval architect of 1800 trying to imagine future vessels. Although he might have some vague ideas about "floating cities," what are the chances that he could have envisioned one of today's giant passenger ships, such as the *Queen Elizabeth II*? It would take a very fertile imagination, and tremendous luck, to predict all-metal boats, petroleum-based and nuclear power systems, refrigerators and air conditioners, navigation and communications equipment, and everything else that comprises today's floating giants.

Like those of physical scientists and engineers, the crystal balls of social scientists grow hazy when we try to peer into the distant future. The engineer's point of departure is present-day technology, and the social scientist's jumping-off point is the collection of political forms, social institutions, cultures, and ways of doing things that surround us today. Between today's planning and tomorrow's emigration humans will change culturally and psychologically, so the first true colonists most likely will be very different from us. Furthermore, the cultural elements that are transported into space will not stay intact, for as emigrants move into new ecological niches, they will evolve new ways. As Jim Dator observes, space is a place to do things that cannot be, or have not yet been, done on Earth.[27] Nonetheless, we have to begin somewhere, and the visions described in this chapter are good places to start.

CHAPTER 14

INTERSTELLAR MIGRATION

Under optimistic projections, we will set forth on the first interstellar voyage before 2100, perhaps within the expected life span of today's children. According to pessimistic projections, for all intents and purposes human interstellar travel is impossible, and the reality is that "you can't get there from here." The problem, given present-day technology, is distance.

Distances between the stars (including our own Sun and its neighbors) are measured in light-years. Since light travels at the speed of 300,000 kilometers per second, one light-year is almost 10 trillion kilometers. (A trillion is a million million.) With the exception of astronomers, who are taught to think big, it is almost impossible for people to grasp the enormity of such distances. Eugene Mallove and Gregory Matloff note that the nearest star, Proxima Centauri, 4.3 light-years from here, is about 273,000 times the distance from Earth to the Sun.[1] If we represent the Sun by a circle with a diameter of 1 centimeter, Earth would be a dot with a diameter of .1 millimeter, located 1 meter away. Pluto would be 42 meters from the Sun, and Proxima Centauri would be 292 kilometers farther out.

STARFLIGHT

The challenge of interstellar travel is covering immense distances within a time frame that keeps the journey meaningful. We have already launched interstellar probes. Pioneer 10 and 11 and Voyager 1 and 2 have completed their work within our solar system and will make their closest approach to other stars as early as 32,600 or as late as 497,000 years from now.[2] Since their primary function is to explore our solar system, they are not taking the most direct route to our nearest neighbor.

If we could travel at the speed of light, then 4.3 years' travel time to Proxima Centauri might not be too bad. Given what we know right now, however, we can't come close to this. Estimated starship speeds are measured in hundredths of the speed of light, and a ship that topped out at .06 times the speed of light would be a snappy performer. Yet because it would take years to accelerate to this speed (and years to decelerate in order to stop at the destination) average speed would be much lower than this. By many estimates, even a "short" interstellar voyage could take centuries. This might not bother an automated probe, but could cause problems for humans.

Destinations

Astronomers have long believed that other solar systems have planets, but only recently have we gained the tools to confirm this. Today's methods enable us to identify only very large planets whose sheer size makes them unlikely places for life. How will we know which stars are likely to have habitable planets where we can live? Will tomorrow's spacefarers risk arriving in a solar system that has only deadly real estate? Although it could be a crapshoot, with continued research we can load the dice in the spacefarers' favor.

In 1964, Steven Dole estimated that about 1 in 200 planets would be habitable; that is, sufficiently similar to Earth as to allow easy settlement by humans.[3] By Martyn Fogg's estimate, this means that we would have to search about 26 light-years to find a planet ready for immediate occupancy.[4] Fogg suggests that in our search for good destinations we should not limit ourselves to such "turnkey," or habitable, planets. We should be open also to biocompatible planets that have the physical parameters for life and can become biospheres after we arrive, and easily terraformable planets that we can make habitable through planetary engineering, perhaps by robots that precede the human settlers. When we combine habitable, biocompatible, and easily terraformable planets, our chances improve. Fogg's model predicts that among 100,000 stars we should find 210 habitable planets, 2,640 biocompatible planets, and 2,692 easily terraformable planets. The implication is that whereas 1 in 476 stars possesses habitable planets, 1 in 37 stars possesses habitable, biocompatible, or easily terraformable planets, a ratio thirteen times higher than for habitable planets alone.

Raymond Halyard hopes for "water planets," ones that contain water, preferably on the surface.[5] He adds that ideally we would find planets that have only one sun, are slightly more than one-half as large but less than twice as large as Earth, and have moderate air temperature, so that the water is not converted into steam. He believes that two advances in astronomy will help us choose good destinations. One is better observational techniques that will help us find and understand planets in nearby solar systems. The other is mathematical models that will allow us to make educated guesses as to which stars are the most promising. Between improved observational techniques, a growing database, and mathematical models, the people who set forth on interstellar missions should have a reasonably accurate idea of what awaits. Perhaps future space crews will be able to access planetographic data on a spherical TV.[6] Robot probes could be sent to check things out, but could significantly delay human departure, since the probes themselves could take decades, perhaps centuries, to complete their scouting mission.

Interstellar Spacecraft

Detailed planning for interstellar spacecraft has been under way since before the first orbital flights. Over the years, many different designs have been brought forward. Some of these are based upon optimistic predictions about future science and technology, but others involve either known technologies and construction techniques or ones that should soon become available. Like airplane, ship, and automobile designs, starship plans reflect trade-offs among size, fuel requirements, and speed.

Rockets such as the Saturn V used in the Apollo program have immense power, but unless we are willing to settle for near-star flybys that will take thousands of years, chemical rockets lack the strength to do the job. Let us take a brief look at some of the alternatives reviewed by Mallove and Matloff and by John Mauldin.[7] I won't even scratch the surface of the technical details; what matters to us is how different starships' performance characteristics are likely to affect their human occupants.

Without question, nuclear rockets are the most promising near-term, advanced propulsion systems. Since the first studies of nuclear propulsion were undertaken in the mid-1940s, states James Dewar, we have learned a lot about how nuclear fuel offers a "compact, portable, inexhaustible, and enormously powerful source of energy capable of overcoming the limits imposed on chemical systems by the laws of physics."[8] In the 1950s, some scientists sought to redirect research from chemical rockets, with their severe weight restrictions and limited ranges, to nuclear rockets that could be constructed like battleships and have sufficient power to deliver massive payloads over vast distances. The 1960s saw major investments in developing plans and prototypes. In 1970, NASA planned to use a rocket known as NERVA (nuclear energy for rocket vehicle applications) for robot and crewed missions in Earth and lunar orbits and for missions to Mars.

Nuclear rockets are of two basic types: fission and fusion. The most common plan for fission rockets calls for nuclear reactors to heat gasses to a very high temperature; the exhausts released from the back of the craft push it forward. The core of the reactor may be solid, liquid, or gaseous. The liquid and gaseous core reactors create successively higher temperatures and thus increase the strength and speed of the exhaust. Another type of fission rocket, the Orion, was designed to explode a rapid succession of tiny A-bombs about sixty meters behind the starship to propel it forward. The basic principles behind fission rockets are well understood, and we could start building a fission rocket today.

Fusion rockets are potential outgrowths of efforts to develop fusion reactions for generating electrical power. Magnetic fields serve as invisible containment vessels or bottles and focus and direct plasma (hot gas) from one end of the bottle. This controlled leak becomes the exhaust that propels the starship. According to Halyard, fusion reactions produce about four times as much energy and twice the exhaust velocity of fission reactions, which could cut interstellar travel time in half. He recommends a version known as a pulsed fusion propulsion system (PFPS) that would use an electron beam or laser to detonate a stream of small pellets of fissionable fuel.[9] Whereas an Orion would be propelled forward by a string of atomic bomblets, the PFPS would be propelled forward by a string of hydrogen bomblets, and at a much higher rate of speed. Fission or fusion, the fuel costs would be staggering for all but a tiny automated probe. On some missions, a short radioactive half-life would cause the fuel to lose much of its "punch" before the ship reached its destination. According to Marc Millis, a nuclear fission shuttle would use a billion supertankers of propellant to reach a neighboring star in 900 years, and a fusion rocket shuttle would use about a thousand supertankers for the same journey.[10]

After a tremendous investment, plans for nuclear rockets were all but abandoned in the early 1970s. This happened, according to Dewar, because of negative public opinion.[11] In 1945, the nuclear age had dawned with the blast of three atomic bombs, two of which demolished cities and killed tens of thousands of people. Then, we spent the next few decades living under the threat of nuclear annihilation during the Cold War. Nuclear power was feared rather than admired, and politicians believed that nuclear rockets were unacceptable to the American people. Promising nuclear rockets were shelved for many reasons, including NASA's fear of the public's fear of radioactivity and a stringent interpretation of the Nuclear Test Ban Treaty of 1963.

An alternative is to use an external power source that can reduce the total mass of the starcraft because there is no need to carry fuel onboard. The potential result is a faster spaceship. This is embodied in the Bussard ramjet that uses a huge scoop—constructed of force fields, not metal—to collect fuel en route. Although we talk about the "vacuum of space," a truly massive scoop might be able to capture enough hydrogen molecules to power this hypothetical interstellar craft.

In the last chapter, I mentioned that Marshall Savage hopes to bounce a laser beam off the back of a spacecraft to help it achieve orbital speeds.[12] In the case of interstellar travel, the laser would require a huge beam on the order of 150 kilometers in diameter. The laser, located in our solar system, would be powered by solar energy collected from the Sun. Voyagers would have to count on human society to be sufficiently stable to maintain the laser's operation over the hundreds of years that it could take for the craft to be pushed to its destination. Such a craft might be more than a little difficult to control because of real-time communication delays. If, a mere two light-years out, the crew determined that it was necessary to redirect the laser to change course, it would take two years for the request to reach Earth and a minimum of another two years for the readjusted beam to alter the direction of the spacecraft.

Still another option is to deploy solar sails near the Sun. After they unfolded, they would be able to accelerate due to the pressure of photons acting upon their truly immense but ultrathin, gossamer surface. Travel by solar sails would be relatively leisurely, at least compared to the other alternatives reviewed here. Marshall Savage suggests

that solar sails might be sent on ahead, carrying freight, while interstellar colonists depart later on a nuclear "clipper" and then catch up.[13]
There are many ways to design hybrid missions that combine various
propulsion techniques; Savage's idea of combining an electromagnetic
launcher with laser propulsion is a good example of this.

MULTIGENERATION MISSIONS

We can divide interstellar missions into two categories: multigeneration
and single generation. All starship designs that use known or reasonably demonstrable technologies will require multigeneration missions,
but some, known as slowships, will require more generations to travel
to a given star than will others, known as fastships. As for single-
generation interstellar missions, many space scientists would say that
we have only hazy or wishful ideas about how to accomplish them, if
they are possible at all.

Slowships

Given that interstellar missions are going to be multigeneration anyway, perhaps we should sacrifice speed for comfort. The idea behind
the interstellar ark, or "worldship," is simple: send the crew forth in a
sluggish luxury liner rather than something that is faster but less habitable. As early as 1929, J. D. Bernal envisioned such a craft made of
materials stripped from planets or asteroids and capable of sustaining
life in comfort.[14]

A worldship is a large settlement or colony that would be sent lum-
bering on its way from one solar system to the next. It would be self-
sufficient, although occasionally work parties might sally forth to replenish raw materials by mining an asteroid or comet. All of the
necessities and conveniences would be there. These include a stable
sociopolitical system, abundant supplies, a variety of workshops and
repair facilities, school systems and hospitals, theaters, and gourmet
restaurants.

The tens or hundreds of thousands of people onboard would have
sufficient space for individual and community activities. They would
have many opportunities for finding friends and marital partners. Living on a worldship would be neither more difficult nor easier than
living in a settlement of comparable size in Earth's orbit. It's just that

some immense propulsion system would be pushing it toward another star. If our descendants choose the slowship option, they will have to be willing to accept incredible engineering challenges to assure the comfort of interstellar migrants.

Fastships

Halyard argues that it will be possible, within this century, to initiate an interstellar mission traveling by nuclear fusion rocket.[15] It would take at least ten years to build the spacecraft and assemble the propellant in low Earth orbit. His proposed spacecraft looks something like a huge flashlight, with the "head" consisting of a thick shield to protect occupants from meteors. A thin shell encloses a combination habitat and landing vehicle, the propulsion system, fuel tanks, and storage areas. The fully self-contained life support system would keep successive generations of space travelers alive during flight and sustain them for a few years after touchdown.

Under this scenario a crew of eighteen would consist of roughly equal numbers of men and women and of three age groups: older couples to serve as leaders and mentors, young adults who would soon start their own families, and the children who eventually would become the last generation among the travelers to have seen Earth. Birth control would be imperative, and genetic engineering or other procedures could be applied to assure healthy offspring and gender balance.

Despite the small size of the crew, it would have many resources at its disposal. Between now and the proposed time of departure there will be tremendous advances in computers, robotics, and nanotechnology, or atom-by-atom construction and repair.[16] Voyagers could have a massively powerful computer that stores all of the information available on Earth. Perhaps there would be a "tool room" capable of repairing and even fabricating worn-out parts, using the materials from the broken ones to do this. As pointed out in chapter 8, helpful but not particularly intelligent robots might perform dangerous and routine functions, and androids that approach (but do not exceed) humans in intelligence might serve as adjunct crewmembers. They would have no life support needs and could be assigned very dangerous tasks.

The crew would live in a lander reminiscent of an overgrown shuttle. Halyard recommends a pleasing interior not unlike that of a contemporary cabin cruiser or motor home—lots of white fiberglass with wood accents. It would include common areas for people to gather, enjoy meals, and watch videos. Each couple would have a tiny bed-room, and there would be a nursery for the children. The travelers would enjoy the usual vegetarian space diet, supplemented with occasional fish or rabbit.

If, upon their arrival, the destination failed to meet minimum habitability requirements, the pioneers would "cease having children" and radio their observations back to Earth to help perfect the models that would keep future pioneers from repeating their mistakes. If all went well, the lander would become the first habitat on the planet. A nuclear power system would be deployed, and the settlers would erect inflatable greenhouses. There, the first colony would grow, achieving a population of about 150 in about seventy-five years. At that point, some settlers would set forth to establish other colonies on the same planet. Halyard reasons that any individual settlement could be beset by a fatal disaster, but it is unlikely that several settlements would succumb to adverse forces at once.

John Mauldin and Edward Regis Jr., as well as Halyard himself, have identified some of the psychological and social pressures that could affect starfarers.[17] Although an interest in science and a dedication to extending humanity beyond our solar system might certainly prompt some people to enlist, the voyagers would not be scientists delighting in their work and achievements. Rather, they would be maintenance technicians in charge of highly automated craft. Except for the final generation, there would be no sense of completion, and generations would die not knowing if the mission would succeed. There would be no latitude for people who have a change of heart, and no chance to return to Earth alive. What happens, for example, if between the time of departure and time of arrival Earth-based astronomers conclude that the destination is totally unsuitable?

Voyagers would lose touch with their past. This is not simply a question of prolonged separation from family and friends: it is a question of losing contact with them completely. Cousins and siblings on Earth would die, but twenty-five light-years out the starfarers would never hear about it. During the early years of migration personal communications would be possible, but these would become slower and less frequent, not just because of increased distances but because tardy responses would discourage people from communicating.

Onboard, there would be limited opportunities for friendships and constrained mating choices. Two men and two women do not each have two choices; once one man and one woman had "found each other" the remaining man and woman would get what's left. Soured relationships could be insufferable, and if someone were unlucky enough to lose his or her partner there might be no replacement. As starfarers became old and could no longer participate actively in the mission, they would have to be cared for by others.

No one would coerce anybody else to set forth on such a flight. However, children who born during the mission would have no choice. It would be, notes Edward Regis Jr., as if the spacefarers had locked up these younger generations and then thrown away the key.[18] Earth would not be known to them, at least not directly; nor for many generations to come would interstellar voyagers know the comfort, spaciousness, and opportunity of orbiting settlements or planetary bases.

Parents would have to instill a strong mission orientation in their children. The primacy of the mission would have to override everything else, and it would have to be so strong that each generation would pass it on to the next. Detailed knowledge of Earth could elicit such longing that at some point this information would be expunged from the data banks. Certain critical skills, such as those for landing and setting up the initial settlement, would have to be passed on from generation to generation and then executed flawlessly on demand. Would computer simulations work well enough? Mauldin wonders what would happen if children rebelled against their parents. One generation might decide to turn back, but another might reverse this decision and pursue their grandparents' goals.

Is it morally permissible to consign future generations to this? Regis points out that *nobody* has control over where he or she is born; whereas some people are born to privilege, others are born into poverty or arrive on Earth during a time of famine or war. "At birth," he states, "all of us are born into conditions that we have not consented to and which may be impossible to escape. In this sense, Earth and the spacecraft are on a par."[19]

The slowship scenario is attractive because it requires less imaginative technology and offers greater comfort for the crew. People who set forth on worldships could sustain themselves under high-quality living conditions over the many generations that they would be en route. The population of a slowship would be large enough to withstand social perturbations such as might be caused by an explosion that killed several of their numbers. If we want to make a multigeneration mission bearable, we should design a worldship so that "getting there is half the fun."

There remain, however, two strong arguments in favor of fast-ships. First, if we insist on waiting for worldships with their staggering habitability requirements, we may never leave our solar system at all. Second, voyagers who arrive via slowships could be in for a nasty surprise. Arriving at their destination, they might discover a thriving human community. They could be beat out by settlers who departed much later, but who, because they had advanced technology, were able to make the journey in much less time. One way around this problem would be to "reserve" destinations on a first departed, first served basis.

SINGLE-GENERATION MISSIONS

Because even efficient starships traveling at a decent fraction of the speed of light could require many generations to reach their destinations, we hope for other techniques that would make it possible for spacefarers who leave for an interstellar destination to be the ones who actually arrive there. One possibility is to invoke warp drives, wormholes, or other time-savers to speed the travelers on their way. Another is to make very long trips survivable by, for example, extending the starfarers' life spans or keeping them in suspended animation for the duration of the flight.

Shorten the Flight

Faster spacecraft is the most obvious strategy for shortening the flight. Interstellar dragsters might be propelled by combining matter and anti-matter, using the energy released in the reaction as the source of power. Right now, antimatter is extremely difficult and expensive to produce, not to mention to store. (Onboard a starship, it would be contained in powerful electromagnetic fields.) A mission to our nearest neighbor would require only ten railroad tank cars of antimatter for propellant.[20]

The faster we go, the more difficult it becomes to achieve even higher speeds. As we come closer and closer to the speed of light, the ship becomes heavier and heavier. The heavier the spaceship, the more powerful the engine and the larger the amount of fuel that is consumed. The more powerful the engine and the more fuel, the heavier the spaceship. This occurs in endless succession.

However, if we could accelerate the ship to relativistic speeds (an appreciable fraction of the speed of light) the crew would gain an advantage. At relativistic speeds, time would pass at different rates inside and outside of the craft. For purposes of interstellar travel this would be a blessing for the people inside. While outside the craft the years would roll by at a ferocious clip, inside the craft time would barely pass. Everything would seem normal to both insiders and outsiders. However, if an outsider could peer in, he or she would make a startling discovery—everything inside the ship would seem to be occurring in slow motion: the spacefarers' activities, the movement of the hands of the clock, the aging process, everything. Over the years, the outsider, who would show the ravages of time, might be annoyed to discover that the men riding inside the craft hadn't even developed a five o'clock shadow. Similarly, from the perspective of the traveler, everyone outside the craft would appear to age very quickly, as if the process were captured by time-lapse photography. Thus, as the light-years ticked off, only a small amount of time would seem to have passed within the starship. The spacefarers would arrive at their destination as if the lengthy trip took only a few months.

The starships that we need for interstellar travel, notes Marc Millis, are the kind that gets us there as quickly and comfortably as those that we see in science fiction movies.[21] This requires abilities that we are not sure we can develop: exceeding the speed of light and manipulating space-time coupling. Erstwhile star travelers need to find ways to capitalize on energy that is available in space itself (thus relieving themselves of the need for railroad cars full of antimatter). This does not mean simply scooping up fuel on the run, as in the case of the Bussard ramjet. It means looking at physics in new and controversial ways, perhaps trying to harness the power of gravity and other force fields.

An alternative to going faster is finding a shortcut. This is the idea behind "space warps" and "wormholes." These are, in essence, tunnels within our universe, perhaps through some hyperdimensional space, that connect locations that we consider to be separated by great distances. By passing through one of these tunnels we would take a shortcut to a location that might be galaxies away. Wormholes might be naturally occurring. Or, through the use of exotic matter (which has negative energy and is quantum mechanical in nature as compared to ordinary matter that has positive energy) and the proper application of electric or magnetic fields we might be able to build one. This is the idea behind the "stargate" in the movie 2001: A Space Odyssey and the contraption built in the movie Contact so that the heroine, played by actress Jodie Foster, could instantly travel several light-years to interact, more or less, with "E.T."

According to theorizing about zero point field energy, there may be enough energy in a cubic centimeter of vacuum to boil away Earth's oceans. The reason for this, note Haisch, Rueda, and Puthoff, is that a vacuum devoid of all matter (stable particles) is a seething quantum "sea of activity" where subatomic particles "flicker in and out of existence."[22] The zero point field involves electromagnetic energy, and since electromagnetic energy is the basis for much of modern technology, manipulating the zero point field could lead to a new form of propulsion. Haisch and his colleagues review studies showing that the possibility of extracting energy from the vacuum does not violate laws of thermodynamics or the theory of relativity, and they cite certain experimental results that are consistent with theoretical predictions.

Despite promising calculations, nobody has located a wormhole or used zero point field energy to propel a vehicle from point to point. Yet in the NASA Breakthrough Propulsion Program, and in privately funded programs such as that sponsored by the International Space Sciences Organization, these and other ideas are on the table with the hope that theorists and experimentalists will take a closer look at them. The goal, states NASA's Millis, is to do credible work on incredible topics.[23]

Lengthen Life

The alternative to speeding up the trip is to find techniques to make it more manageable for the voyagers. Perhaps through advances in biotechnology we could extend life three thousand years or so. If we could do this, a few centuries in flight might not seem that long. Essentially, we are preprogrammed to wear out within a particular time frame. There is tremendous variability in the lengths of individual lives, and for really old people the last few years may not be particularly rewarding. Perhaps we could learn how to refresh the genes and activate the mechanisms that keep the body both alive and in good repair.

Spacefarers would be checked carefully for good health prior to departure, and because of the closed ecology there would be no reason to concern ourselves that a new disease might appear onboard. Small robots might be able to perform operations and thus handle such maladies as appendicitis, gallstones, or a slipped disk. Maybe a nanotechnology-based chemical "tool room" could manufacture fresh pharmaceuticals of any type. There is always the risk of accidents, and the cumulative probability of accidental death sometime over hundreds or thousands of years would be much greater than for the seventy to eighty years that most of us are allotted. The risks could be reduced by great caution on the part of people whose most likely method of dying was accidentally.

Another alternative is "suspended animation" that halts metabolic processes and forces people into a trance- or dreamlike state that eases the passage of many years. Somehow they would be reawakened a short time before reaching their destination and, judging by science fiction portrayals, very little time would pass between waking up and regaining peak fitness. Imaginary spacefarers who sleep large portions of their lives away never seem to suffer from muscle atrophy, locked joints, and other terrible problems that devastate patients kept immobile in hospitals on Earth. Perhaps some sort of exercise machine built into their travel couches could keep them fit by moving their limbs or applying electric current.

Drugs and hypnotic states would leave starfarers requiring intravenous feeding, catheters, and other types of intensive life support. Better yet might be cryonics suspension, or freezing. The spacefarer's body temperature would be lowered maybe four hundred degrees and he or she would be kept "on ice" for most of the voyage. As the spacecraft approached the destination, an automated system would thaw and revive the crew in time to supervise landing operations. At present, we can freeze sperm, skin, and other tissue, but this must be done very carefully so that the cells don't rupture. However, we are a long way from being able to revive frozen people.

Part of the challenge is that different temperature gradients and chemicals are used for freezing, storing, and thawing different types of human tissue, and the procedures that preserve some cells kill others. People who have themselves frozen in liquid hydrogen immediately after death are betting on two major advances in medicine: techniques for thawing entire bodies in such a way that they are not further damaged, and cures for the illnesses that killed them in the first place.

INTERSTELLAR HUMANITY

Perhaps, before the end of this millennium, humans will be spread throughout our solar system and will emerge as a presence near neighboring stars. Perhaps there will be huge orbiting communities, a giant Moonbase, and a thriving society on Mars. Perhaps small groups of adventurous pioneers will mine comets in the Oort Cloud. Whatever the actual time line and sequence of events, if and when space settlement is under way the frontier will move erratically but inexorably outward.

If we could observe human progress up close we might see some huge worldships carrying populations entering their tenth generation moving lazily from one solar system to another, while other explorers use breakthrough propulsion systems to dart here and there. The size of human communities might range from twenty-five or so people to huge extraterrestrial societies measured in the billions. Some people might live on barren rocks that contain thousands of tons of valuable minerals, while others inhabit planets that are more lush than Earth. After all, with countless planets to choose from, our descendants might find some that are far more inviting than our current home.

What would humanity be like as it spread among the stars? Although this pushes the limits of reasonable extrapolation, human migrations on Earth may help us understand human migrations in space. Many of the best ideas on the subject were spawned at a conference held at Los Alamos National Laboratories in the early 1980s and described in Ben R. Finney and Eric M. Jones's book *Interstellar Migration and the Human Experience*.[24] Like tomorrow's migrants will do, yesterday's migrants completed lengthy and dangerous voyages and lost contact with their home communities. Some spent generations isolated from all other societies. The limitation of these analogues is that most of these voyages took place centuries ago by foot, horseback, and canoe. Polynesians, Vikings, and ancient Greeks did not have the advantage of androids, robots, expert systems, genetic engineering, nanotechnology and other technologies (either real, or so far only imagined) that we hope will help spacefarers on their way.

Population

Because travel within the solar system will be so much easier than travel to the stars, we will be able to send wave after wave of settlers to places like the Moon and Mars. Within our solar system, population will reflect not only the size of the founding colony (the first group of settlers to arrive) and their birthrate (traditional or artificial) but also the arrival of new groups of immigrants.

At least initially, we will not be able to send additional personnel to extrasolar destinations, so the size of the population will depend on the number of people who arrive in the founding settlement and their birthrate alone. The larger the founding colony and the higher the birthrate, the quicker the community will gain biological and economic viability. We want an initial community that is large enough to (1) ensure a steady stream of offspring free of defects due to inbreeding, (2) survive heavy casualties, and (3) provide the economic base for abundant and high-quality goods and services.

According to some analyses, even very small founding settlements (perhaps as few as ten people) can constitute a successful breeding population. Inbreeding is unlikely to yield genetic defects, providing that the founders come from drastically different genetic populations. J. B. Birdsell describes a man known as "Afghan" who, after mating with his Australian aborigine wife, mated with his daughter and his daughter's daughter, ultimately fathering great-grandsons who were seven-eighths his genetic composition.[25] None of Afghan's descendants showed harmful effects of inbreeding. More typically, anthropologists recommend a breeding population of approximately five hundred people to minimize the risks of inbreeding. It is not necessary that they all live together; they could live successfully in bands of twenty-five but should get together with other bands for purposes of mate selection.

Large populations are required in order to support high-quality life. William A. Hodges points out that the emigrants could live in relative comfort aboard the starship but would begin feeling the pinch as the population grew after arriving at the destination.[26] Suppose that on-board the starcraft there were plenty of computers, but that the pioneers lacked the materials and means to construct more. This means that computers would become a rarity as soon as the first native-born settlers sought access to them.

As long as the population remained small, there would not be enough people to justify specialty and luxury goods. At first, people would have to be generalists rather than specialists and limit production to necessities. As the community grew, it would support craftsmen and craftswomen by forming a large enough market for the fine goods that they could produce. Also as the population grew, specialists would appear. Persons who served as pharmacist-veterinarian-dentist-physician would be superseded by pharmacists, veterinarians, dentists, and physicians. Later on, general practitioners would be supplemented by medical specialists such as internists and radiologists. According to

Hodges, an extrasolar community would require a million people to achieve the quality of life that is enjoyed within the solar system. Of course, we can always hope that new technology will compensate for the small size of the population.

As the population grew, explorers would set off from one settlement to found another. Many of these communities would be located at vast distances from one another, and it would not be surprising if different groups of settlers were to lose touch. As people adapted to new environments that differed radically from Earth's they could evolve into very different biological forms. James Valentine predicted that within 2 million years "the descendants of *Homo sapiens,* scattered across the Galaxy, would exhibit a diversity of form and adaptation that would astound us residents of Earth today."[27] Yet, this need not take millions of years, for these changes could be facilitated by genetic engineering. Sometimes, perhaps, humans from one part of the galaxy would find it impossible to mate successfully with humans from another part. They would no longer belong to the same biological group: speciation will have occurred.

Cultures

Studies of terrestrial migrations give clues for understanding cultures in outer space.[28] Perhaps the most important determinant of the new culture will be the culture that the settlers thought they left behind. Those from England will drink synthetic tea, and those from the United States will drink synthetic coffee. Personal motivation, of course, will play a role: we might expect cultural differences in groups of settlers who emigrate to find increased personal freedom, as compared to those who leave Earth to tap the vast riches of the galaxy.

Elements of a home culture tend to be perpetuated in a group's new location, with certain exceptions. When a group is fleeing a political regime we would expect a different political system in the new community. Certain events that occur prior to departure or en route might influence the new culture. The leader's personality might have an effect: for generations, he or she might be venerated and citizens might ask themselves, "Is this how Captain Noah would have done it?" The group's experiences, especially when it is struggling to stay alive, would be important. Travelers who are running low on resources but who then stumble across an asteroid ripe for mining could develop a culture in which asteroids gain ritual significance. Or, a person who performs a heroic service such as sacrificing his or her life so that others can live could become a cultural icon.

At first, settlers are likely to be somewhat authoritarian in their outlook. Strong leadership and close adherence to group norms could help the group through its first years of hardship. Reaffirmation of belief in the group's fundamental correctness would validate the initial decision for emigrating from Earth and constitute rationalization that would suppress pangs of disappointment that might occur when they see their new home close up. Rituals from the past might persist, but in attenuated or stripped-down form. One example might be a simplified criminal justice system, with abbreviated proceedings reminiscent of a court martial rather than the elaborate proceedings that typify Anglo-American jurisprudence (see chapter 13). Religious rituals might be salient for those who maintain their earlier beliefs or create new ones, because like all rituals they can promote group solidarity, and group solidarity is essential for survival. Yet the rituals themselves could be simplified if there are not enough people and artifacts to permit performing them with their customary frills and flourishes.

At time of departure from Earth, terrans and the spacefarers will have very similar cultures. From then on, the two cultures will diverge. New developments on Earth will not be readily available to people in interstellar flight. At first, it will be possible for terrans and spacefarers to exchange ideas, but this will become increasingly difficult as a function of growing real-time communication delays. The drift will increase over successive generations as those who had experience on Earth die and as terrestrial events become less and less relevant to the immediate needs of the spacefarers.

Spacefarers may embark with the latest and best technology, but it will be impossible for them to stay at the cutting edge. We would expect the eventual rise of indigenous technologies based on the needs and resources available in space. Right now, it is difficult to imagine people showing greater technological achievement in the undeveloped environment of space than in the highly developed confines of Earth. At some point, however, there may be more people living off of our planet than on it, and the expatriates may have the edge.

What would happen if we were to find a planet so hospitable to life that it is already inhabited? The chances of this are remote.[29] Nonetheless, Raymond Halyard suggests that if the planet were inhabited we might do well to find a location that separated us from the natives.[30] For example, we might avoid land masses inhabited by intelligent creatures and settle instead on islands not yet reached by intelligent life. Over a thousand years or so, we would studiously avoid contact with their civilization, allowing it to grow and develop in its own way. Later on, we might initiate relationships, but we would do so slowly, perhaps over a period of thousands of years.

Who would have the intellectual edge? The best guess that we can make right now is that we humans are of average intelligence. Although we might like to think of ourselves as front-runners, as did Europeans upon arriving in the New World, there is at best only a fifty-fifty chance that we would have the upper hand.

Lessons from history suggest that contact between two previously isolated civilizations has met with, at best, mixed success. We are most familiar with the dislocations, pain, and misery resulting when a group with advanced weapons comes across a culture ill-equipped to defend itself. Generally, the technologically advanced group has prevailed, sometimes on the basis of military action. Other empires were built on promises and gifts rather than hostile military action. All that was requested was that the local authorities acknowledge and perhaps pay tribute to the "great sovereign," for which they would in return receive protection from third parties and be allowed to carry on more or less as they pleased.

According to recent calculations, the average civilization in our galaxy could be more than *1 billion years* farther down the evolutionary path than is our own.[31] If this is so, then it could be impossible for representatives of the two cultures to relate to one another. Most likely there would be signs of such a civilization—for example, patterns of energy use—that starfarers could detect several light-years out. If this occurred, the starfarers might choose to change course for an alternative destination.

Interstellar Politics

At first, communities in space will follow the laws and customs of their home societies. We might expect a fair degree of coordination among different space settlements and Mother Earth. Initially, this would come about because the different groups of settlers would arrive in space with similar laws and similar ways of doing things, and because UN or other terrestrial authorities would try to impose order upon them. However, as the number of settlements increases, and as they become spread out over vast interstellar distances, it would become increasingly difficult for any centralized authority to maintain control. The problems of surveillance, transportation, and (in the case of interstellar distances) real-time communication would make it very difficult to impose political uniformity. There will be no Galactic Empire operating under a unified system of command.

Different groups of settlers themselves might seek coordination in certain areas that yield mutual benefits. There could be agreements on interstellar communication and transportation, with shared use of Earth's airwaves and seas providing the model. Similarly, there could be agreement to collaborate in certain areas such as assisting one another in the event of a large meteor strike or other disaster. There might also be health regulations designed to prevent the transmission of disease from one colony to another, and collaborative ventures on the part of communities in close proximity; for example, construction of a starship to travel to a neighboring solar system. Although space colonies would evolve and sharpen their identities, we might expect some coordination, at least among neighbors.

Declan O'Donnell proposes a "Metanation" modeled loosely on the United Nations to encourage synergy between "humankind" located on Earth and "spacekind" located everywhere else.[32] The nations of Earth (all nations, not just the rich ones that have taken the lead in space exploration) would work together to provide the economic and legal infrastructure to support people in space. Spacekind would abide by laws that discouraged mutual interference and encouraged cooperation in some areas but also made it possible for them to reap direct economic benefits from their work. O'Donnell's goal is to make emigration into space an orderly process and to transform the abstract principle that space is the "common heritage of mankind" (CHOM) into something meaningful and workable.

To these ends, O'Donnell identifies potentially useful terrestrial models. These include the Tennessee Valley Authority and the New York Port Authority. Under the latter model, which could be applied on the Moon or Mars, an administrative group, chartered by the UN or other terrestrial authorities, maintains order and provides settlers with core services such as a spaceport and water supplies. It also ensures the rights of people from all nations on Earth to come and go freely and engage in free enterprise. O'Donnell and his associates at the Denver-based United Societies in Space have petitioned the Colorado courts to form a Lunar Economic Development Authority.

For several reasons, including growing recognition that it is methodologically ineffective as well as morally repugnant, war should be a rarity.[33] If events unfold as predicted by those who expect wide-scale emigration from Earth, people will live under conditions that are conducive to peace: they will be well-to-do materially, and they will have control over their own destinies. Furthermore, mathematical models show that, simplistic views of "survival of the fittest" not withstanding, it is the peaceable rather than the warlike societies that are likely to survive, at least when they enter into collective security agreements.[34] Despite its salience in our minds, war has always been the exception rather than the rule, and, on Earth, war is becoming less common.[35]

If nothing else, notes Michael Hart, having plenty of elbow room for population growth, coupled with vast distances among interstellar colonies, will discourage war. Although it may be aggressive people who occupy interstellar space, they will be so spread out that they will have very little contact with one another, and, if interstellar voyages are measured in centuries, military invasions may become unpopular. "Under such circumstances," states Hart, "a typical civilization might be involved in war only once every 50,000 years, perhaps much longer."[36]

CONCLUSION

Will humans as a species survive long enough to migrate to the stars? On that distant day, what sorts of technology will be available? One thing is certain: in the absence of practical worldships or major breakthroughs in advanced propulsion systems, such trips will be truly formidable. The work of the NASA Breakthrough Propulsion Program suggests that new technologies may be possible. It is easy to scoff, because such technologies are not available right now. Still, as Haisch, Rueda, and Puthoff point out, only fifty years ago "the concept of space travel was regarded by most (including scientists who should have known better) as science fiction"; and only slightly more than fifty years before that many people had the mindset that, for humans heavier than air, flight was impossible. "We have," they write, "come to a new millennium and the first glimmerings of how to go about finding a way to achieve interstellar travel have started to appear on the horizon."[37]

Maybe we can populate the stars while remaining at home. One possibility, suggested by Frank Tipler, begins with replicating the contents of human minds in powerful computers.[38] After all, both computers and brains are information-processing devices, and we are rapidly approaching that time when computers have the human brain's information-processing capacity. Then, it would be the computer representation of the person—which Tipler treats as identical to the person in terms of a sense of self and continuity, memories, perceptions, emotions, and other psychological qualities—that makes the voyage. Long after your physical body had perished, the computer representation of your mind would move peacefully in space, in a deep, dreamless sleep or amused by virtual reality—a representation that we hope would not be erased by the stray electromagnetic pulse. The beauty of this, if it worked, is that there would be no need for life support as we normally think of it. Upon arrival you would be reactivated, and perhaps your mind would be reconstituted in the computer "brain" of an android. We will leave it to technologists to handle the details, and to philosophers to decide whether or not it would be the real "you" resurrected at the destination.

A less exciting but perhaps more workable alternative is to send genetic codes from Earth along with media for nurturing life once the spacecraft has reached its destination.[39] Upon arrival, the genetic instructions would be carried out, and life would begin to develop. The process could begin with the evolution of single-celled organisms whose metabolic processes would begin to terraform the extrasolar planet. Perhaps over many millennia intelligent life-forms would appear. The upshot is that although humans themselves would not travel interstellar distances, we still could seed life throughout the galaxy. There are many conceivable paths to interstellar migration, and the ones that we actually will tread (if any), remain to be seen.

CHAPTER 15

RESTORING THE DREAM

Optimistic discussions of our future in space have an inspirational quality and stress the inevitability of migration from Earth. Even as yesterday's settlers drove Conestoga wagons across the American plains, tomorrow's will drive spaceships to the Moon, to Mars, and to the asteroid belt. The time line is uncertain, but some of these discussions are couched in terms of decades, not centuries. If we ourselves don't live long enough to get there, then at least our children will have a chance to be the first humans on Mars.

How inexorable is our drive to reach the stars? To get there, we need know-how and wealth. Right now, we may occupy a narrow window where we have both the technology and the resources to proceed. If we delay too long, this technology might be used for war or other destructive purposes. If we are wasteful, if we let our oil wells run dry and strip all of our mines of ore, we could be stranded on Earth.

There is another way that we could lose momentum. We don't expect the shuttle to remain in operation beyond 2015, and there are no government-sanctioned plans for human spaceflight beyond the ISS. Somewhere in the second or third decade of this new millennium we could lose the cadre of professional spacefarers who should lead us into space. Our experiential base will wither. To regain the practical knowledge that we need, we may have to restart the cycle; that is, begin anew with brief orbital missions. Just as the loss of important people and key information has made it infinitely more difficult to assemble a Saturn V rocket in 1999 than in 1969, two or three decades from now it may be more difficult to assemble a crew than it is today.[1]

What Went Wrong on the Way to the Future?

Judging by predictions made over the past four decades, as of right now shuttles should have undertaken hundreds of flights; the ISS should have been operational for years; we should have returned to the Moon; and we should be on the verge of setting off for Mars, if not already settled there. In actuality the excitement and promise of the 1960s space program is a faint memory, and many predictions about how spacefaring would be in the year 2000 have been wrong.[2] Instead of a continuing series of firsts, we must content ourselves with repetitive shuttle and space station missions. The pleas of scientists and entrepreneurs who have thoughtful plans for accelerating our progress into space fall on deaf ears. To some observers, the space program is drifting if not moribund.

Exactly how "stuck" are we at the forty-year mark? Although some earlier projections now seem fanciful, it is not fair to say that space exploration is in full retreat. Since 1960, science, government, and industry have strengthened the infrastructure that we need to move into space. This strengthening includes advances in metallurgy and materials science, computers and information-processing technology, and above all, miniaturization that lowers the cost of lifting payloads into orbit. Propulsion systems have improved, as have life support, navigation, and communication systems. We have increased our understanding of the solar system and are learning more about other stars. All of this knowledge is useful for leaving our cradle, Earth.

Communications satellites are a multi-billion-dollar industry. We are beginning to do wonders with satellite remote sensing and navigation systems. Despite significant birthing problems, the Hubble space telescope has returned over a hundred thousand images, and in the late 1990s we witnessed some magnificent automated missions, such as the landing of a rover on Mars.

Since astronauts and cosmonauts first orbited Earth, we have gained substantial experience living in space. Since Apollo, the number of people who have been in space has grown about tenfold. We learned from Apollo-Soyuz, Skylab, numerous Salyuts, and Mir, as well as from scores of shuttle flights. Delayed or not, the ISS will further improve our understanding of the human side of spaceflight and help us prepare the next generation of crews.

Perhaps we can draw an analogy between erecting a building and developing a space program. To the untrained observer, the building's progress is uneven. Once the foundation is done, the framework and then shell are rapidly completed. Sidewalk engineers and prospective tenants rejoice. Then, progress slows to a snail's pace. Yet, hidden inside the shell, workers are installing water and electrical lines, erecting interior walls, building floors, and then painting. Conspicuous initial progress creates unrealistic, optimistic expectations that draw attention away from the massive amounts of work still needed to complete the building.

The rapid advances of the 1960s space program are analogous to the speedy construction of a frame and shell. Early progress obscured all of the work remaining before any humans would be ready to permanently occupy space. Perhaps future historians will recognize the past forty years of space missions as an orderly and coherent progression on the long march into space, rather than as the period of slow and erratic steps that it seems to be right now.

Still, space enthusiasts are restless. They find it tempting to lash out at weak political and industrial leadership, an ill-informed and apathetic public, decaying government bureaucracies, greedy corporations that put exorbitant profits before the common good, and failures to promote the "right" technology. Although it is very easy to latch onto one of these culprits, no single cause slows our progress into space. The question "Whatever happened on the way to the future?" must be addressed at three levels: the broad context of public opinion, the interests and goals of the many constituencies that are affected by space exploration, and the functioning of specific organizations such as NASA.

Public Opinion

The enthusiasm of the American public for space exploration peaked at the time of the first landing on the Moon, and public opinion has been evenly divided ever since.[3] About half of the people polled support a space program. While this support is widespread, it is also "soft." Many people who profess support for space exploration don't want to help pay for it. Many people lack enthusiasm because they know that they will never get there personally. For most of us, spaceflight is reserved for people who are "not like me." In this sense, aviation and space have developed in very different ways. In *Space and the American Imagination*, Howard E. McCurdy notes that at its inception the commercial airline industry fostered a campaign promising "Anyone can fly."[4] People were bombarded with photos of happy passengers, and one of the reasons that the industry shifted from flight stewards to flight stewardesses in the 1930s was to underscore the ease of flying for women.[5] In the early days of airplanes, people could take five-dollar barnstorming rides, and today, anyone who has a few hundred dollars can fly around the world. During the early years of spaceflight, on the other hand, spacefarers were very special, almost superhuman people. More recently, the occasional paying customer might orbit Earth on Mir, but with "ticket prices" in tens of millions of dollars the experience isn't for everyone. If and when it arrives, space tourism—which will make space travel "for someone like me"—may renew public interest in space and generate enthusiasm for all aspects of space exploration.

The lack of drama in the contemporary space program may also have had a chilling effect. Many people who were enthusiastic about the succession of dramatic firsts during the 1960s lost interest as the space program encountered difficulties and became back-page news. Leonard David suspects that the "barnstorming" of the early years contributed to high public support of that space program.[6] According to David, everything since the Moon has been anticlimactic. To move forward again we must restore human drama—"No Buck Rogers, no bucks." But alas, there is a Catch-22 here, because it's equally true that "No bucks, no Buck Rogers." Recent accounts of life aboard Mir and the dangerous space walks required to build the ISS are returning high drama to the high frontier.

McCurdy advances the thesis that a growing discrepancy between expectations and realities may have extinguished enthusiasm.[7] In the late 1940s, before NASA was formed, scientists like Hermann Oberth and Wernher von Braun sought to engage public support for spaceflight. This involved, in part, drawing on powerful images already present in American popular culture. As a result of such efforts, space travel became continuous with Columbus setting forth for the new world, cowboys in the Wild West, and Amundsen and Scott fighting their way to the South Pole.

In the 1950s, McCurdy writes, television specials and a series of articles in then-popular magazines such as Collier's helped make space travel seem imminent and real. Audiences learned that although their construction would be challenging, huge spaceships intended for the Moon or Mars could be based on foreseeable technology. Great artists of the day prepared lavish illustrations that captured public imagination, much as preceding generations of artists had generated interest in the Old West. NASA continued this trend, writes McCurdy, by creating an "aura of competence" around the Apollo program. The costs and dangers were relegated to the background while stories about the "gee whiz" technology and "can do" attitudes saturated the media.

The high expectations created by marketing were not matched by subsequent realities. Space travel was dangerous, expensive, and unavailable on a wholesale basis. The kinds of spacecraft and missions that caught America's attention were not the kinds that Congress could fund. Enthusiasm, driven by the imagination, dwindled once the realities became clear.

Finally, there is the problem of competing priorities. Right now, the public must weigh real and immediate cost against gains that, if they materialize at all, will do so in the very distant future. Most people, perhaps the vast majority, are concerned about immediate, practical issues. For them, spaceflight is visionary and impractical. Space exploration seems particularly irrelevant for citizens of underdeveloped countries, whose income may not exceed three hundred dollars a year and who live at a subsistence level. Although their descendants may be among the greatest beneficiaries in some future era of space commercialization, this is small consolation if the present population does not have enough to eat.

We won't go to the Moon or Mars as a result of a popular vote, but increased political support would help. The first step is developing a better product: a space program that engages the public's attention, is not hopelessly thwarted by astronomical costs, and reliably fulfills its promises. The second step is informational campaigns intended to recapture the public's interest. An informational campaign should focus on key audiences, including influential people who are known as "opinion leaders," on representatives of the media, and on younger generations, including children.[8] Roger Handberg points out that the ISS reflects the dreams of the current generation and does not necessarily reflect the dreams of future generations.[9] Only if we can capture our children's interests can we assure continued progress into space.

Space advocacy groups could play a potent role in winning public support and acting as lobbyists in support of space activities. In 1984 Michael A. G. Michaud found that many such groups were small and unstable; despite the potential for unification by an overriding enthusiasm for space, they were not even loosely coordinated with one another.[10] Different groups had competing agendas. On occasion different groups formed coalitions, but later on these would disintegrate. Oftentimes politics within and between advocacy groups worked against effective lobbying.

Since Michaud's study the situation may have improved somewhat. Today, there are many space advocacy groups, including broad-based groups such as Carl Sagan's Planetary Society and the National Space Society. There are also special interest groups, such as the Artemis Project, dedicated to reestablishing our presence on the Moon; Robert Zubrin's Mars Society, dedicated to the Mars Direct program; and the small Colorado-based United Societies in Space, committed to establishing political and legal frameworks to promote an orderly movement into space. Moreover, there are powerful industry groups and at least one association of space explorers. The National Space Society acts as an umbrella that encompasses many different groups while still allowing them to retain their unique identities and special interests. With continued growth and convergence, space advocacy groups could accelerate human progress in space.

Constituencies

The space program, states W. D. Kay, is influenced by governmental organizations, scientific groups, industrial concerns, and many other constituencies whose interests NASA must take into account.[11] While we may like to think that NASA executives should "make no compromises," they cannot always follow this principle in a world where organizations are forced to reach agreements, form alliances, and overcome opposition. NASA's attempts to come to grips with these different constituencies leads to concessions and compromises that undercut the Space Agency's preferred plans.

Initially, NASA was built around the National Advisory Committee for Aeronautics, and then subsumed the Army Ballistic Missile Arsenal consisting of Wernher von Braun and his German rocket-science team, along with elements of the air force and navy that were involved in missile development. NASA co-opted these different groups and transformed them, if not into a unified whole, then at least into a collection of people who would sink or swim together. Spurred by the Cold War, Congress made huge appropriations. Since there were very few competitors for space dollars and almost no knowledgeable critics, NASA was, in essence, a well-funded monopoly subject to few checks and balances.

Shortly after this, the White House and Congress hired staff experts and increased their oversight of NASA. No longer was NASA free to make plans solely on its own initiative. Proposals had to navigate a sea of organizations and agencies, including the executive and legislative branches of government, the military, space industries, environmental protectionists, and advocacy groups. The plot thickens in the case of international programs, which must reach agreements with international partners, who in turn must please their own multiple constituencies.

Every 179 years, the outer planets of our solar system—Jupiter, Saturn, Uranus, and Neptune—are ideally aligned for exploration by robotic craft. Since the days of the American Revolution, notes D. Rubashkin, this alignment occurred only twice: in the presidencies of Thomas Jefferson and Jimmy Carter.[12] In the early 1970s, NASA sought to launch such a robot mission. This "grand tour" would have been very expensive to develop, in part because it would have required a reliable self-diagnosing and repairing computer that would remain operational for 5 billion kilometers and ten years.

The president, Congress, and the Office of Management and Budget considered the project too expensive. Scientists were unenthusiastic because the mission would have drained funds away from a space telescope (ultimately the Hubble) and other projects they held dear. Advocates of human space exploration worried that the "grand tour" would divert too much money from piloted spaceflight. Ultimately, the NASA administrator killed the project to keep human space exploration alive. As a low-cost compromise, separate probes were launched toward a few of the outer planets. We will have our next chance for a "grand tour" in only 150 years, or sooner with new technology.

W. D. Kay's point is that the more constituencies that get into the act, the more concessions and compromises will shape a final product. In the course of trying to appease if not satisfy all of these different constituencies, the technology becomes overburdened. Costs increase, programs run behind, and a series of "make do" decisions leads to a disappointing product.

A superordinate goal—a goal that is of overriding importance to several constituencies—would help. This requires structuring missions in ways that a variety of groups would find something of real value. They must *want* the mission to succeed and not merely compromise with NASA in order to reach their own separatist ends.

Organizational Dynamics

In his book *Inside NASA*, Howard E. McCurdy points out that no organization can sustain the excitement and growth of NASA's first few years.[13] At first, the glamour of space, abundant funding, and minimal red tape enabled NASA to hire the "best and brightest" and made it possible for employees to work on projects of their choosing. In times of affluence and growth, organizations can afford to move down multiple tracks (for example, developing new Moon and Mars missions at the same time) and "bury" their mistakes. Yet even in the early 1960s it became impossible for NASA to do everything itself, and it began hiring outside contractors. The collaboration worked, and the overall success of programs through Apollo is well known.

When the Apollo era drew to a close, funds were tight and outside contractors were playing greater roles. NASA officials spent less time doing technical work and more time currying political support and overseeing contracts. Within NASA, the proportion of managers rose and the proportion of scientists and technicians declined. Thus, although still successful in many ways, an organization formerly known for its technical orientation and hands-on approach became known for its bureaucratic qualities.

In 1992, when McCurdy was completing this particular study, the field of management was undergoing a revolution. Efficiency and quality became overriding goals. Executives sought to strengthen values, eliminate or at least reduce bureaucracy, simplify procedures, thin the ranks of middle management, dump nonessential activities, and refocus on the organization's central purposes (also known as "getting back to the knitting"). Slavish adherence to past practices and bureaucratic rules were to give way to innovation and entrepreneurship.

The NASA administrator Dan Goldin, who arrived fresh from industry that year, was clearly aware of this revolution, and, reciting the mantra "faster, better, cheaper," tried to address NASA's problems. He took visible steps to reduce the bureaucracy and increase the amount of science. Over the next few years, this became evident in NASA initiatives in astrobiology, planet imaging, and breakthrough propulsion physics. According to Joan Lisa Bromberg, Goldin accelerated an earlier trend toward creating stronger partnerships with business.[14] Operation of the shuttle, for example, was transferred to a space industry consortium, and efforts were undertaken to develop new spacecraft funded partly by NASA and partly by private corporations.

Under Goldin's plan, NASA would remain involved for a few years in each new venture (such as the ISS and Moon or Mars bases). When each of these became a routine operation, it would be turned over to other operators, including private corporations. At that time NASA itself would raise its sights to the next frontier. This is not to imply that, under Goldin, NASA experienced an unblemished string of successes. The ISS fell farther and farther behind schedule, some of the NASA-industry partnerships fell through, and cheaper spacecraft were not necessarily better. This is to say that he attempted to move a giant bureaucracy into a brave new era and, we hope, set the stage for further success.

BACK TO THE FUTURE

Each day civilization comes to numerous choice points. Futurists believe that the pattern of choices made determines our future. Today, as we branch off in one direction or another, we *invent* tomorrow. Years ago, our choices poised us for an exciting future in space: supposedly one of adventure and unlimited personal, economic, and social opportunity. These choices, reflected in proposals for interplanetary voyages and space settlements, were not confirmed by subsequent choices, and we veered off in other directions. How do we get "back to the future"—the future that we imagined two or three decades ago, a future that includes large numbers of workers in orbit and a continuing human presence on the Moon and Mars?

In his book *The Space Station Decision,* McCurdy argues that we must begin with a clear vision and definite goals.[15] He points out that President John F. Kennedy's command to land on the Moon by the end of the 1960s was a loud and unequivocal mandate that cleared the path for one of the greatest scientific and technological accomplishments of all time. There was no clear vision for building a space station. Although NASA had hoped to launch a space station as early as the 1960s, for a long time it was unable to develop a plan that gained strong presidential and public support. In the early 1980s, NASA was able to convince President Ronald Reagan to support planning for a space station. In an attempt to make the project attractive, NASA gave very low cost estimates that omitted contingency funds and "extras" (such as the cost of placing the station in orbit) that many people would consider part of the package. As the project progressed the real cost became increasingly evident. To help offset objections to the huge budget, NASA repeatedly redefined the space station, diminishing both its size and purpose. In this way, NASA attempted to get the support that it needed on a year-by-year basis. Since NASA had not been able to build a long-term commitment to it, the space station was an attractive target for politicians who sought to reduce the federal budget. Rather than trying to earn acceptance of the ultimate goal—a fully crewed space station that could support an array of scientific and commercial activities—NASA muddled along from year to year, conducting study after study and vainly striving to make the final costs seem less enormous. There was no consistent vision and there was no congressional buy in.

In *Living and Working in Space*, Philip R. Harris argues that setting and reaching goals to establish a continuing human presence in space is best achieved through massive collaboration.[16] It will require concerted action on the part of physical, biological, and social scientists; government and industry; technologists and humanists; management and labor; different groups of space advocates; and people from all nations, not just those currently spacefaring.

Repeatedly, we have proven our ability to bring macroprojects to completion. Such projects of varying magnitude include construction of Egyptian and South American pyramids, completion of the transcontinental railroad in the United States, the Suez and Panama Canals, the Tennessee Valley Authority, the Manhattan Project that led to the development of the atomic bomb, the Apollo program, and the Chunnel connecting England and France. Like establishing ourselves on the Moon or Mars, these were daring projects that stretched our imagination and involved high risk, tremendous financial commitment, and in many cases, state-of-the-art or even emerging technologies. Before they are completed, people tend to see these projects as too expensive if not impossible.

Planning such projects requires a broad perspective, in-depth knowledge, creativity, and "outside the box" thinking. Participants have to be creative, flexible, open to feedback, and predisposed to reach consensus. Such macroplanning rests on a strong and clear vision for the future. It brings together many different factions and constituencies and stimulates their enthusiasm so that common interests override differences. The plans themselves set forth goals and priorities, identify the necessary technologies along with feasibility analyses, and above all, identify realistic funding mechanisms. Necessarily the plans would be flexible and incomplete, and it is a foregone conclusion that despite widespread buy in not everyone would accept them.

At periodic intervals, the U.S. government has developed comprehensive plans for humans in space. Two such efforts were accomplished by the National Commission on Space, chaired by the former NASA administrator Tom Paine during the Ronald Reagan presidency, and the Space Exploration Initiative undertaken during the George Bush administration.[17] The National Commission viewed the shuttle as the lead partner in space exploration. Increased orbital activity would develop an infrastructure for a return to the Moon. In the course of this, we would develop the technology and knowledge to establish a continuing presence on Mars. The National Commission's work was eclipsed by the *Challenger* explosion. The Space Exploration Initiative was thwarted by exorbitant projected costs and a lack of presidential follow-through. Thus, beyond the ISS, there are no clear priorities for humans in space.

In contrast to McCurdy and Harris, B. Alex Howerton contends that large-scale planning efforts may actually retard our progress.[18] Grandiose plans take too long to formulate and may fail by virtue of their wishful nature and complexity. Too much effort goes into planning, not enough effort into completing the actual job. Considering both planning and execution, macroplans have long time lines and thereby contain the seeds for their own destruction.

North Americans, and to a lesser extent western Europeans, tend to be impatient. We can concentrate, but only for short periods of time. For example, the Gulf War focused attention because it lasted only a few days. The 1996 capture of accused murderer O. J. Simpson on the Los Angeles freeways was even more riveting: it was only a cocktail or two after the networks' initial telecasting of the pursuit before Simpson was apprehended by the police. Unlike the Chinese, who are influenced by their expectations for future centuries, contemporary Westerners are such that it is difficult to gain sustained interest in long-term ventures. Even if we could gain initial support for a long-term plan, public opinion and political priorities would change, and support would dwindle before the program reached fruition.

Howerton's alternative calls for a series of inexpensive steps that we can take in a timely manner. Rather than making spaceflight more complex and expensive, we should make it simpler, cheaper, and more accessible. He finds the development of computers instructive. At first, computers were extremely large, difficult to operate, and useful only to the government and a handful of big businesses and laboratories. Now they are mass produced, easy to use, and priced within the reach of almost everyone.

Another possibility is to set large goals, such as lunar industrialization, but then reach them in a series of small, profitable steps. Haym Benroya proposes a Lunar Development Corporation that would involve financiers, managers, lawyers, scientists, and engineers.[19] All activities would be "for profit," but it would take about twenty years from initiation to the first year of colonization. For this reason, it is imperative to find ways to earn profits along the way. This could involve the development of "dual use" technologies that have immediate application on Earth as well as future application on the Moon. We can visualize the process as something like trying to go from Los Angeles to New York by train. There are many paths and stops, and the challenge is to find ways to make each stop profitable. This evolves into two questions: *"Can we work our way across the country?* and *Which path puts us in the best financial and technological condition at the end of the trip?"* We need to answer parallel questions before returning to the Moon or embarking for Mars.

Cutting Costs

While it is all well and good to compare the cost of a piloted expedition to Mars to the cost of Apollo, or to the huge amounts of money that have been spent for purposes of defense, price tags in the range of $50 to $500 billion give us pause. We may feel better if we prorate such costs over a decade or two, or express them as a tiny percentage of a burgeoning gross national product. Nonetheless, billions are billions, and any money poured into space is money that is not available for other purposes.

Without a doubt, the best single way to speed our progress into space is to decrease the price of getting there. One way to reduce the cost is to demand accountability. For years, much of the work done by aerospace and other contractors was done on a cost-plus basis. This meant that the payments they received were based on their actual costs in terms of equipment, material, and labor, plus an additional percentage for profit. This arrangement structured the situation in such a way that contractors who ran up big bills by buying lots of expensive material and hiring hordes of workers made much more money than their efficient, frugal competitors. Business is done differently now, but there is still opportunity for abuse. We must be wary of funding formulas that seek gargantuan sums for optional consultants, unnecessary travel, infrastructure development, and overhead.

Another strategy is to eliminate needlessly expensive technology. Certainly we must insist upon high quality and reliable equipment, and we do need some redundancy. Yet we should also keep our eyes open for cheaper alternatives. During the early U.S. space program, considerable money was spent developing a ballpoint pen that would work under conditions of microgravity, where there was no gravity to draw the ink downward. It works great, but the Russians cut costs by giving their cosmonauts pencils. The search for inexpensive, alternative technologies is one of the themes in Marshall Savage's plan to "colonize the Universe in eight easy steps."[20]

The greatest challenge is finding less expensive ways to lift material and people into orbit. If we can reduce launch costs from a few thousand dollars to a few hundred dollars per pound we will jump the greatest hurdle to space settlement. Low-cost lifts to orbit are the closest thing to a "magic bullet" for accelerating our progress into space. The most promising alternatives right now are fully reusable spaceplanes and single-stage-to-orbit rockets.

At the "Case for Mars VI Conference" held in Boulder, Colorado, in 1996, everyone, it seemed, had his or her own pet way to reduce the cost of getting to Mars. These included big, heavy rockets that were cheap and simple to construct but that would burn immense amounts of chemicals, and small, lightweight rockets that were expensive and complicated but fuel-efficient. Some advocates extolled the wonders of rockets that got their start as they popped up to the ocean's surface, and other advocates urged spacecraft that left Earth on the back of a jumbo jet.

Some proposals seemed rooted more firmly in their inventor's belief system than in known technology. The same was true of early aircraft, but early aircraft, notes Freeman Dyson, were allowed to evolve.[21] Early newsreels include footage of airplanes with six sets of wings, ornithopters with flapping wings, and craft with propellers pointing straight up to provide lift, all either failing to get off the ground or destroying themselves shortly afterward. Some of these radical designs worked, but not very well or at exorbitant cost. Because airplanes were tested experimentally, hundreds of designs evolved into a few basic configurations that reign over our skies today. Unfortunately, spaceships cannot be evaluated by the law of survival of the fittest, because they are too expensive. We have to rely on computer testing to perfect a very limited number of designs at any one point in time. As in the case of the first orbiting space shuttle, the first real test (as compared to simulated tests) may come when the first Mars ship departs with its crew.

Partnerships

Space exploration and development can be made more manageable through sharing the costs. This partnering could involve different nations, government and industry, or consortia of private interests.

International collaboration is the basis for the European Space Agency and made possible the ISS. While international collaboration may be necessary, and in certain ways desirable, it also has its drawbacks. International ventures are difficult to organize and administer. Treaties and agreements are not always honored. Internal politics and economic difficulties may make it difficult for a nation to fulfill its obligations. It may be necessary to make sacrifices and reach compromises to maintain the interest of international partners. In some cases, promoting spaceflight as an "international" venture adds to rather than decreases instability.[22]

Another form of partnership involves the public and private sectors. The question "Should the government or private industry fund human movement into space?" might be rephrased. The more fruitful question is, "What are the roles of the public and private sectors as we move into space?"

At first, not only did the U.S. government take the initiative in space exploration, it co-opted other players, for example, by contracting work to outside vendors. Now, NASA does not have enough money to take the next steps, some of which are big ones. The corporations best equipped to lead us are working on defense contracts and are preoccupied with nonpiloted missions, such as launching telecommunications satellites. "For NASA," writes Roger Handberg, "which defines itself emotionally and organizationally in terms of crewed flight, there is need but no money, whereas the private sector probably sees no need as they presently perceive the situation."[23]

Howerton believes that the U.S. government has never had that much interest in spaceflight per se.[24] Instead, it has pursued spaceflight for other purposes, such as to enhance national prestige and defense. Competing priorities and weak or inconsistent public support mean that we cannot count on the government, by itself, to initiate and sustain a continuous human presence in space. Furthermore, a contracting system that involves high overhead and the view that space activities are expenditures rather than investments tends to perpetuate high costs. Supporting space exploration through the private sector requires assembling large numbers of investors and combining corporate might. However, for many potential investors, space is not all that attractive right now.

Because of the high cost factor, truly immense investments are required for solar power satellites, lunar bases, space hotels and the like to reach fruition. Many conventional investments require only modest sums of money. Moreover, many conventional investments yield immediate dividends. Unless we can follow Benroya's strategy of finding profitable stops along the way, it will take too long to realize a return on the investment. Most investors do not envy the position of colleagues who invested in the Chunnel between England and France. Eventually, the Chunnel will become immensely profitable, but at that point it is likely to be the investors' children and grandchildren who will reap the rewards. For most hard-nosed, economically driven investors (and they are the only ones with plenty of money to invest) there are more attractive investment opportunities on Earth.

Investors require large returns to justify the high costs and risks. Some products from space (such as space power, crystals, and pharmaceuticals) have to compete against comparable products developed on Earth. We cannot get by forever with showcase products that are heavily subsidized or purchased at a premium price because of their glamorous origin. We must develop markets for unique products that are produced under conditions of weightlessness, and open new markets by moving into potentially profitable areas, such as space tourism. In some future era, space settlers can provide goods and services for one another. People operating mass drivers on the Moon may be able to send raw materials to orbiting colonists cheaper than competitors who are trying to export raw materials from Earth. But the reality is that these markets don't exist right now.

In addition to the normal risks associated with new ventures, competition with terrestrial products, and a fickle marketplace, space investors are forced to use risky technology. A payload could explode before it gets into orbit, taking a large fortune with it. Insurance companies may offer coverage, but because the technology is risky the premiums are very high. When spaceflight becomes routine some of these risks will decline, in part as a result of improving technology and in part because the insurers will be able to prorate risk over scores and then hundreds of flights, most of which will not experience significant problems.

Investors must be sensitive also to legal risks. According to a beautifully framed document in my office, I hold title to a specific star. According to this document, my title was registered in Switzerland. Although my retirement plans do not call for the arrival of an interstellar craft bearing tribute, some interesting questions arise. Will the grandchildren of today's titleholders demand to be paid off before mining operations can begin on "family property" on the Moon? Although we treat such deeds as novelties and jokes, their legal status is an important question for people who otherwise might invest in lunar operations.

One of the greatest concerns, according to Declan O'Donnell and Philip R. Harris, is a patchwork quilt of treaties and agreements that leave many ambiguities concerning the rightful use of space resources.[25] The United States' position has been that one can simply go to space, establish a base, and start mining. Yet the Moon Treaty of 1979— which the United States did not sign—consigns the resources of space to "the common heritage of mankind" (CHOM). This means that everyone, from spacefaring and nonspacefaring nations alike, is to share in the benefits. But how should we interpret this? Does this mean that no individual nation is allowed to use space resources? That these must forever remain untouched until some sort of benefit-sharing is arranged? Does this mean that the spacefaring nations are free to use these resources but must pay other nations royalties or licensing fees? Who would get what, and could any nation demand up-front payment? Such ambiguities add to the overall risk of space enterprise.

CHOM is only one legal issue that needs to be resolved before investing massive amounts of funds in space. If a failing missile or reentering satellite causes damage, those who launched it are held responsible. On the other hand, if someone else's space junk wrecks your orbiting hotel, it could be very difficult to collect damages. We need to rationalize space laws and treaties to promote an orderly, peaceful, and prosperous movement beyond Earth.

Alan Wasser's approach to a public-private partnership would reinstate the land-grant system so that investors who agreed to develop lunar or planetary resources would be given huge tracts of land for subsequent development.[26] U.S. railroads during the nineteenth century were encouraged to develop new lines by offers of massive parcels of government land, extending perhaps five or ten miles on each side of the tract. Not only did they receive free land for the railroad itself, but also land for building towns, farming and mining operations, and other activities that would generate income in their own right as well as create additional products for the railroad to transport. A one-hundred-square-kilometer strip mine on the Moon might qualify operators for an additional package the size of Alaska. This could be of irregular shape, but the land itself would have to be contiguous and relatively compact so as to leave some opportunity for latecomers. Thus, early investors would have rights to land that would bring even greater rewards as the mining and manufacturing activities expanded. If the corporation quit the land, then it could be reassigned to another group of entrepreneurs. New legal mechanisms would have to be in place to support a successful land-grant program.

CONCLUSION

Richard P. McBrien divides the past fifty thousand years into 800 sixty-two-year lifetimes.[27] In less than one of these lifetimes we have progressed from launching satellites not much larger than basketballs to brief orbital missions, trips to the Moon, a partially reusable space-plane, and an almost continuously occupied space station, Mir. Within this same lifetime, which commenced in 1959, we may be able to return to the Moon and take our first steps on Mars.

We seek to establish a continuing human presence in space to conduct science and to pursue knowledge, to tap the fabulous resources of the universe, to prosper psychologically, and to renew society. Technology and people must work hand in hand, but this book has focused on the human role. We considered human factors broadly defined to extend beyond biomedical adaptation and the human-machine interface, and to include personality, social relationships, and the broad political and organizational contexts for human activity in space. Increasing crew size, increasing diversity of crew composition, increasing mission duration, and burgeoning technology prompt us to rethink the human side of spaceflight.

Spacefarers withstand a formidable list of dangers and hardships: high acceleration, life-threatening radiation, microgravity with all of its attendant problems, austere if not primitive living conditions, isolation from families and friends, and confinement within a cramped microsociety. It hasn't always been easy, but spacefarers have met the challenges thus far. By selecting people who are appropriate for the job, training them properly, and giving them the right habitats, equipment, supplies, and procedures we have provided the support they need to conquer and tame space.

In less than a dozen years we progressed from Sputnik to the first human beings on the Moon. Although not always the stuff of newspaper headlines, progress has been made ever since. The shuttle is a reliable if expensive mechanism for transporting people to and from space, and, despite "losing" the Moon race, the Russians moved steadily forward, keeping a series of space stations occupied and setting record after record for human endurance in orbit. We know much more about spacecraft construction and propulsion technology, life support systems, and biomedical adaptation than we did when the space era began. Growing experience in space itself, coupled with research in spaceflight-analogous environments, has expanded our understanding of the human side of spaceflight. In order to assure our continued success in space, we must address human issues the same way we address technical issues: with foresight, planning, and careful research.

Early efforts to "sell" the space program created high expectations that have not been matched by recent progress. With the advantage of twenty-twenty hindsight, it is tempting to heap scorn upon those we believe have delayed our entry into space. Space enthusiasts bemoan a do-nothing Congress that can't grasp the significance of humankind's next evolutionary step, blame a national space administration that never seemed to make the transition from the unlimited wealth of the 1960s to the relative austerity that followed, accuse greedy contractors who can't contain costs, condemn inept bureaucrats everywhere, and of course, castigate an uninformed public that doesn't realize that untold riches lie just beyond their grasp. If we are to accelerate progress, we must dramatically reduce the cost of launching materials into orbit. We can do this through eliminating waste, selecting low-cost alternatives when they are readily available, and above all, developing efficient new technologies. We must find ways of making investment in space more attractive, develop new partnerships, and reduce legal risks.

In the twenty-first century, visitors to Washington's Smithsonian Air and Space Museum can tour a replica of Skylab. At Johnson Space Center, tourists can see a huge Saturn V rocket sitting like a giant lawn ornament, a testimony to wonders past. On the Moon, there is discarded equipment of no further use and astronauts' footprints frozen forever in time. Far above Earth's surface are satellites that are integral to Earth's central nervous system. The ISS is there along with Soyuz, Mir and the shuttle. On Earth, NASA-contracted educators in unmarked vans drive from school to school demonstrating space suits and teaching with Moon rocks. No longer individual gladiators representing their own countries, astronauts and cosmonauts train together. Businesses ranging from tiny design studios and engineering firms to huge aerospace conglomerates contemplate ventures in space. As defense contracts decrease and the heavens become saturated with communications satellites, aerospace giants may think more seriously about new ventures such as space manufacturing and tourism. Science fiction writers, illustrators, and movie producers are still at work. Part memory, part reality, and part dream of the future, "the space program" is alive.

Notes

1. WHY SPACE?

1. Bryan Burrough, *Dragonfly: NASA and the Crisis aboard Mir* (New York: HarperCollins, 1998).

2. William K. Douglas, "Psychological and Sociological Aspects of Manned Spaceflight," in *From Antarctica to Outer Space: Life in Isolation and Confinement,* ed. Albert A. Harrison, Yvonne A. Clearwater, and Christopher P. McKay (New York: Springer-Verlag, 1991), pp. 81–89.

3. John S. Lewis, *Mining the Sky* (Reading, Mass.: Helix Books, 1997).

4. Daniel S. Goldin, "Steps to Mars" (paper presented at the "Steps to Mars Conference," July 25, 1995).

5. Paul S. Hardesen, *The Case for Space: Who Benefits from Space Exploration on the Last Frontier?* (Shrewsbury, Mass.: ATL Press, 1997); B. Alexander Howerton, *Free Space: Real Alternatives for Reaching Outer Space* (Port Townsend, Wash.: Loompanics Unlimited, 1995).

6. R. C. Parkinson, "Review of Rationales for Space Activity," *Journal of the British Interplanetary Society* 51 (1990): 275–80.

7. Paul D. Lowman Jr., "T Plus Twenty-Five Years: A Defense of the Apollo Program," *Journal of the British Interplanetary Society* 46 (1996): 71–79; David M. Harland, *Exploring the Moon: The Apollo Expeditions* (Chichester, U.K.: Springer-Praxis, 1999).

8. Bruce Cordell and Joan Miller, "Young People and Rationales for Human Mars Missions," in *The Case for Mars IV: Considerations for Sending Humans,* ed. Thomas R. Meyer (San Diego, Calif.: American Astronautical Society/Univelt, 1997), pp. 407–24.

9. Http://www.challenger.org/, October 25, 1997.

10. Cordell and Miller, "Young People and Rationales for Human Mars Missions."

11. Hardesen, *The Case for Space.*

12. Scott Sacknoff and Leonard David, *The Space Publications Guide to Space Careers* (Bethesda, Md.: Space Publications, 1998).

13. Hardesen, *The Case for Space.*

14. Jesco Von Puttkamer, "Space Humanization: Always a Mission to Planet Earth," *Space Governance* 1 (1994): 18–27.

15. William H. Siegfried, "Return to the Moon: A Commercial Program to Benefit Earth" (paper presented at the 10th International Space Plans and Policies Symposium, Space Exploration and Development—International Aspects, Beijing, China, October 1996).

16. Ibid.

17. Lewis, *Mining the Sky;* Sanders D. Rosenberg, "Lunar Resource Utilization," *Journal of the British Interplanetary Society* 50 (1997): 337–52.

18. Lewis, *Mining the Sky.*

19. Gregory L. Matloff, "The Near-Earth Asteroids: Our Next Planetary Destinations," *Journal of the British Interplanetary Society* 51 (1998): 267–74.

20. Gerard K. O'Neill, *The High Frontier* (New York: Bantam Books, 1978); Marshall Savage, *The Millennial Project: Colonizing the Galaxy in Eight Easy Steps* (Boston: Little, Brown, 1994).

21. David Ashford and Patrick Collins, *Your Spaceflight Manual: How You Could Be a Tourist in Space in 20 Years* (London: Headline Books, 1990).

22. Ibid.

23. Marsha Freeman, *How We Got to the Moon: The Story of the German Space Pioneers* (Washington, D.C.: 21st Century Science Associates, 1993).

24. Marsha Freeman, "Krafft Ehricke's Extraterrestrial Imperative," *Space Governance* 2 (1995): 20–24.

25. Ibid., p. 21.

26. C. M. Hempsell, "History of Space and Limits to Growth," *Journal of the British Interplanetary Society* 51 (1998): 323–36.

27. Ben R. Finney and Eric M. Jones, "The Exploring Animal," in *Interstellar Migration and the Human Experience,* ed. Ben R. Finney and Eric M. Jones (Berkeley and Los Angeles: University of California Press, 1984), pp. 15–25.

28. Ibid., p. 21.

29. Lawrence G. Lemke, "Why Should Humans Explore Space?" in *Strategies for Mars: A Guide to Human Exploration,* ed. Carol R. Stoker and Carter Emmart (San Diego, Calif.: American Astronautical Society/Univelt, 1996), p. 7.

30. Philip R. Harris, *Living and Working in Space: Human Behavior, Culture, and Organization,* 2d ed. (Chichester, U.K.: Wiley-Praxis, 1996).

31. Ibid., p. 3.

32. Sacknoff and David, *The Space Publications Guide to Space Careers.*

33. Walter Cunningham, *The All-American Boys* (New York: Macmillan, 1977), p. 29.

34. Frank White, *The Overview Effect* (Boston: Houghton Mifflin, 1987).

35. Ibid., p. 39.

36. Ibid., p. 26.

37. Ibid., p. 43.

38. Jerry M. Linenger, *Off the Planet* (New York: McGraw Hill, 2000), p. 71.

39. Edgar Mitchell and Dwight Williams, *The Way of the Explorer* (New York: Putnam, 1996).

40. Harrison H. Schmitt, "The Millennium Project," in *Strategies for Mars: A Guide to Human Exploration*, ed. Carol R. Stoker and Carter Emmart (San Diego, Calif.: American Astronautical Society/Univelt, 1996), p. 37.

41. Albert A. Harrison and Joshua Summit, "How Third Force Psychology Might View Humans in Space," *Space Power* 10 (1991): 85–203.

42. Schmitt, "The Millennial Project."

43. Michael Collins, *Liftoff* (New York: Grove Press, 1988).

44. W. D. Kay, *Can Democracies Fly in Space? Revitalizing the US Space Program* (Westport, Conn.: Praeger, 1995).

45. Nathan C. Goldman, "The Mars Base: International Cooperation," in *The Case for Mars II*, ed. Christopher P. McKay (San Diego, Calif.: American Astronautical Society/Univelt, 1985), pp. 65–72.

46. Brian O'Leary, *Exploring Inner and Outer Space: A Scientist's Perspective on Personal and Planetary Transformation* (Berkeley: North Atlantic Books, 1989).

47. Http://shuttle-mir.nasa.gov/, July 22, 1997.

48. Parkinson, "Review of Rationales for Space Activity."

49. Lemke, "Why Should Humans Explore Space?"

2. Spaceflight Human Factors

1. Edgar D. Mitchell and Dwight Williams, *The Way of the Explorer* (New York, Putnam, 1996).

2. Elwyn Edwards, "Introductory Overview," in *Human Factors in Aviation*, ed. Earl L. Wiener and David C. Nagel (San Diego, Calif.: Academic Press, 1988).

3. Philip R. Harris, *Living and Working in Space: Human Behavior, Culture, and Organization*, 2d ed. (Chichester, U.K.: Wiley-Praxis, 1996), p. 49.

4. N. A. Kanas and W. E. Fedderson, "Behavioral, Psychiatric, and Sociological Problems of Long Duration Missions," NASA TM X-58067 (1971).

5. Harris, *Living and Working in Space*, p. 66.

6. Space Studies Board, National Research Council, *A Strategy for Space Biology and Medical Science for the 1980s and 1990s* (Washington, D.C.: National Academy Press, 1987), p. 168.

7. Space Studies Board, National Research Council, *A Strategy for Space Biology and Medical Science for the New Century* (Washington, D.C.: National Academy Press, 1998), p. 195.

8. Patricia A. Santy, *Choosing the Right Stuff: The Psychological Selection of Astronauts and Cosmonauts* (Westport, Conn.: Praeger, 1994).

9. Albert A. Harrison, "On Resistance to the Involvement of Personality and Social Psychologists in the US Space Program," *Journal of Social Behavior and Personality* 1 (1986): 315–24.

10. Howard E. McCurdy, *Space and the American Imagination* (Washington, D.C.: Smithsonian Institution Press, 1997).

11. Ibid.

12. W. D. Kay, *Can Democracies Fly in Space? The Challenge of Revitalizing the US Space Program* (Westport, Conn.: Praeger, 1995).

13. Bryan Burrough, *Dragonfly: NASA and the Crisis aboard Mir* (New York: HarperCollins, 1998), p. 15.

14. Vyascheslav I. Myasnikov and Iltuzar S. Zamaletdinov, "Psychological States and Group Interactions of Crew Members in Flight," in *Space Biology and Medicine,* ed. Arnauld E. Nicogossian, Stanley R. Mohler, Oleg G. Gazenko, and Anatoliy I. Grigoriev, vol. 3, bk. 2 (Reston, Va.: American Institute of Aeronautics and Astronautics, 1996), pp. 419–32; Yuri A. Aleksandrovskiy and Mikhail A. Novikov, "Psychological Praxis and Treatments for Space Crews," in *Space Biology and Medicine,* ed. Arnauld E. Nicogossian, Stanley R. Mohler, Oleg G. Gazenko, and Anatoliy I. Grigoriev, vol. 3, bk. 2 (Reston, Va.: American Institute of Aeronautics and Astronautics, 1996), pp. 433–43.

15. Mary M. Connors, "Exploring Personality: Social Psychology's Role in the US Space Program" (paper presented at the 94th Annual Convention of the American Psychological Association, Washington, D.C., August, 1986).

16. Mary M. Connors, Albert A. Harrison, and Faren R. Akins, *Living Aloft: Human Requirements for Extended Spaceflight,* NASA SP-483 (National Aeronautics and Space Administration, 1985).

17. Mary M. Connors, Albert A. Harrison, and Joshua Summit, "Crew Systems Dynamics," *Behavioral Science* 39 (1994): 183–212.

18. McCurdy, *Space and the American Imagination.*

19. Myasnikov and Zamaletdinov, "Psychological States and Group Interactions of Crew Members in Flight," p. 419.

20. Harvey Wichman, *Human Factors in the Design of Spacecraft* (Stony Brook, N.Y.: Monograph Series of the New Liberal Arts Program, 1992).

21. S. Vinograd, ed., *Studies of Social Dynamics under Isolated Conditions* (Washington, D.C.: Sciences Communication Division, George Washington University Medical Center, 1974).

22. Jack Stuster, *Bold Endeavors: Lessons from Polar and Space Exploration* (Annapolis, Md.: Naval Institute Press, 1996).

23. Albert A. Harrison, Yvonne A. Clearwater, and Christopher P. McKay, eds., *From Antarctica to Outer Space: Life in Isolation and Confinement* (New York: Springer-Verlag, 1991).

24. Ben R. Finney, "Scientists and Seamen," in *From Antarctica to Outer Space: Life in Isolation and Confinement,* ed. Albert A. Harrison, Yvonne A. Clearwater, and Christopher P. McKay (New York: Springer-Verlag, 1991), pp. 89–102.

25. W. Reid Stowe and Albert A. Harrison, "One Thousand Days Nonstop at Sea: Lessons for a Mission to Mars," in *The Case for Mars IV: The International Exploration of Mars — Considerations for Sending Humans,* ed. Thomas R. Meyer American Astronautical Society (San Diego, Calif.: American Astronautical Society/Univelt, 1997), pp. 395–406.

26. Marshall T. Savage, *The Millennial Project: Colonizing the Galaxy in Eight Easy Steps* (Boston: Little, Brown, 1994).

27. Robert L. Helmreich, "Psychological Research in Tektite II," *Man-*

Environment Systems 3 (1973): 125–27; Roger Bakeman and Robert L. Helm-reich, "Cohesiveness and Performance: Covariation and Causality in an Un-dersea Environment," *Journal of Experimental Social Psychology* 11 (1973): 472–89; Jack Stuster, *Space Station Analogues* (Santa Barbara, Calif.: Anacapa Sciences, 1984).

28. Stuster, *Space Station Analogues.*

29. Benjamin B. Weybrew, "Three Decades of Nuclear Submarine Re-search: Implications for Space and Antarctic Research," in *From Antarctica to Outer Space: Life in Isolation and Confinement,* ed. Albert A. Harrison, Yvonne A. Clearwater, and Christopher P. McKay (New York: Springer-Verlag, 1991), pp. 105–14.

30. Ibid.

31. Peter Suedfeld, "Groups in Isolation and Confinement: Environments and Experiences," in *From Antarctica to Outer Space: Life in Isolation and Confinement,* ed. Albert A. Harrison, Yvonne A. Clearwater, and Christopher P. McKay (New York, Springer-Verlag, 1991), pp. 135–46.

32. D. T. Andersen, C. P. McKay, R. A. Wharton Jr., and J. D. Rummell, "An Antarctic Research Outpost as a Model for Planetary Exploration," *Journal of the British Interplanetary Society* 43 (1990): 499–504.

33. Christopher P. McKay, "Antarctica: Lessons for Mars," in *The Case for Mars II,* ed. Christopher P. McKay (San Diego, Calif.: American Astro-nautical Society/Univelt, 1985).

3. HAZARDS AND COUNTERMEASURES

1. Sharon Begley and Adam Rogers, "The Ultimate Thrill Ride," *Newsweek* 54–55 (November 30, 1998).

2. Phillip S. Clark, "Space Debris Incidents Involving Soviet/Russian Launches," *Journal of the British Interplanetary Society* 47 (1994): 379–91.

3. Brian Harvey, *The New Russian Space Programme: From Competition to Cooperation* (Chichester, U.K.: Wiley-Praxis, 1996).

4. Sybil Parker, *Concise Encyclopedia of Science and Technology,* 3d ed. (New York: McGraw-Hill, 1994), p. 1740.

5. This discussion of the moon and planets is based on Bruce Murray, Michael C. Malin, and Ronald Greely, *Earthlike Planets: Surfaces of Mercury, Venus, Earth, Moon, and Mars* (San Francisco: W. H. Freeman and Company, 1981); Anthony R. Curtis, ed., *Space Almanac,* 2d ed. (Houston: Gulf Publishing Company, 1992); and Parker, *Concise Encyclopedia of Science and Technology.*

6. Edgar Mitchell and Dwight Williams, *The Way of the Explorer* (New York: Putnam, 1996).

7. Harvey, *The New Russian Space Programme.*

8. Parker, *Concise Encyclopedia of Science and Technology,* p. 1440.

9. G. Harry Stine, *Living in Space* (New York: M. Evans, 1997), pp. 57–73.

10. Malcolm Ritchie, "General Aviation," in *Human Factors in Aviation,* ed. Earl L. Wiener and David C. Nagel (San Diego, Calif.: Academic Press, 1988), p. 568.

11. Arnauld E. Nicogossian and Donald G. Robbins, "Characteristics of the Space Environment," in *Space Physiology and Medicine,* ed. Arnauld E. Nicogossian, Carolyn Leach Huntoon, and Sam L. Pool, 3d ed. (Philadelphia: Lea and Febiger, 1994), pp. 50–52.

12. Leonard David, "Space: A Fountain of Youth?" *Final Frontier* (September-October 1997): 25–27.

13. Rosalind A. Gryme, Charles E. Wade, and Joan Vernikos, "Biomedical Issues in the Exploration of Mars," in *Strategies for Mars: A Guide to Human Exploration,* ed. Carol R. Stoker and Carter Emmart (San Diego, Calif.: American Astronautical Society/Univelt, 1996), pp. 225–40; Millard F. Reschke, Deborah L. Harm, Donald E. Parker, Gwenn R. Sandoz, Jerry L. Homick, and James M. Vanderploeg, "Neurophysiologic Aspects: Space Motion Sickness," in *Space Physiology and Medicine,* ed. Arnauld E. Nicogossian, Carolyn Leach Huntoon, and Sam L. Pool, 3d ed. (Philadelphia: Lea and Febiger, 1994), pp. 248–60.

14. Reschke et al., "Neurophysiologic Aspects," p. 229.

15. Jeffrey F. Davis, James M. Vanderploeg, Patricia A. Santy, Richard T. Jennings, and Donald F. Stewart, "Space Motion Sickness during 24 Flights of the Space Shuttle," *Aviation, Space, and Environmental Medicine* 46 (1988): 1185–89.

16. Reschke et al., "Neurophysiologic Aspects."

17. Lakshimi Putcha et al., "Pharmaceutical Use by US Astronauts on Space Shuttle Missions," *Aviation, Space, and Environmental Medicine* 70 (1999): 705–8.

18. John B. Charles, Michael W. Bungo, and G. William Fortner, "Cardiopulmonary Function," in *Space Physiology and Medicine,* ed. Arnauld E. Nicogossian, Carolyn Leach Huntoon, and Sam L. Pool, 3d ed. (Philadelphia: Lea and Febiger, 1994), pp. 286–304.

19. Anastasia Toufexis, "The Hazards of Orbital Flight," *Time* (February 28, 1983): 48.

20. John Billingham, "An Overview of Selected Biomedical Aspects of Mars Missions," in *The Case for Mars III—General,* ed. Carol R. Stoker (San Diego, Calif.: American Astronautical Society/Univelt, 1985), pp. 157–70; Grymes, Wade, and Vernikos, "Biomedical Issues in the Exploration of Mars"; M. Mazher Jaweed, "Muscle Structure and Function," in *Space Physiology and Medicine,* ed. Arnauld E. Nicogossian, Carolyn Leach Huntoon, and Sam L. Pool, 3d ed. (Philadelphia: Lea and Febiger, 1994), pp. 317–26; Charles F. Sawin, "Biomedical Investigations Conducted in Support of the Extended Duration Orbiter Medical Project," *Aviation, Space, and Environmental Medicine* 70 (1999): 169–80.

21. Victoria Garshnek, "Exploration of Mars: The Human Aspect," *Journal of the British Interplanetary Society* 43 (1990): 169–80; Jeffrey R. Davis, "Medical Issues for a Mission to Mars," *Aviation, Space, and Environmental Medicine* 70 (1999): 162–68.

22. Victor S. Schneider, Adrian D. LeBlanc, and Linda C. Taggert, "Bone and Mineral Metabolism," in *Space Physiology and Medicine*, ed. Arnauld E. Nicogossian, Carolyn Leach Huntoon, and Sam L. Pool, 3d ed. (Philadelphia: Lea and Febiger, 1994), pp. 327–33.

23. Paul C. Rambault, Carolyn S. Leach, and G. D. Whedon, "A Study of Metabolic Balance in Crewmembers of Skylab IV," *Acta Astronautica* 6 (1979): 1313–22.

24. Charles W. DeRoshia and J. E. Greenleaf, "Performance and Mood-State Parameters during 30-Day 60° Head-Down Bed Rest with Exercise Training," *Aviation, Space, and Environmental Medicine* 64 (1993): 522–27.

25. Carolyn Leach Huntoon, Peggy A. Whitson, and Clarence F. Sams, "Hematologic and Immunologic Functions," in *Space Physiology and Medicine*, ed. Arnauld E. Nicogossian, Carolyn Leach Huntoon, and Sam L. Pool, 3d ed. (Philadelphia: Lea and Febiger, 1994), pp. 351–62.

26. Sawin, "Biomedical Investigations Conducted in Support of the Extended Duration Orbiter Medical Project," 169–77.

27. Ted D. Wade, Philip G. Smaldone, and Richard G. May, "Automation of Fitness Management for Extended Space Missions," in *The Case for Mars III — Technical*, ed. Carol R. Stoker (San Diego, Calif.: American Astronautical Society/Univelt, 1985), pp. 171–88.

28. Rosalind A. Grymes, Charles E. Wade, and Joan Vernikos, "Biomedical Issues in the Exploration of Mars," in *Strategies for Mars: A Guide to Human Exploration*, ed. Carol R. Stoker and Carter Emmart (San Diego, Calif.: American Astronautical Society/Univelt, 1996), pp. 225–39; Donald E. Robbins and Tracy Chui-Hsu Yang, "Radiation and Radiobiology," in *Space Physiology and Medicine*, ed. Arnauld E. Nicogossian, Carolyn Leach Huntoon, and Sam L. Pool, 3d ed. (Philadelphia: Lea and Febiger 1994), pp. 167–93; L. W. Townsend and J. W. Wilson, "The Interplanetary Radiation Environment and Methods to Shield from It," in *Strategies for Mars: A Guide to Human Exploration*, ed. Carol R. Stoker and Carter Emmart (San Diego, Calif.: American Astronautical Society/Univelt 1996), pp. 283–323.

29. Benton C. Clark and Larry W. Mason, "The Radiation Show Stopper to Mars Missions: A Solution," in *The Case for Mars IV: Considerations for Sending Humans*, ed. Thomas R. Meyer (San Diego, Calif.: American Astronautical Society/Univelt 1996), pp. 101–13.

30. Robbins and Yang, "Radiation and Radiobiology."

31. Garshnek, "Exploration of Mars."

32. Townsend and Wilson, "The Interplanetary Radiation Environment and Methods to Shield from It."

33. Stine, *Living in Space*.

34. Sam L. Pool, Arnauld E. Nicogossian, Edward C. Moeseley, John J. Uri, and Larry S. Pepper, "Medical Evaluations for Astronaut Selection and Longitudinal Studies," in *Space Physiology and Medicine*, ed. Arnauld E. Nicogossian, Carolyn Leach Huntoon, and Sam L. Pool, 3d ed. (Philadelphia: Lea and Febiger, 1994), pp. 375–93.

35. Mary M. Connors, Albert A. Harrison, and Faren R. Akins, *Living*

Aloft: Human Requirements for Extended Spaceflight, NASA SP-483 (Washington, D.C., U.S. Government Printing Office, 1985), pp. 26–27.

36. Grymes, Wade, and Vernikos, "Biological Issues in the Exploration of Mars."

37. Roger D. Billica and Richard T. Jennings, "Biomedical Training of US Space Crews," in *Space Physiology and Medicine,* ed. Arnauld E. Nicogossian, Carolyn Leach Huntoon, and Sam L. Pool, 3d ed. (Philadelphia: Lea and Febiger, 1994), pp. 394–401.

38. William J. Crump, Bennie J. Levy, and Roger D. Billica, "A Field Trial of the NASA Telemedicine Instrument Pack in a Family Practice," *Aviation, Space, and Environmental Medicine* 67 (1996): 1080–85.

39. Ibid.

40. Putcha et al., "Pharmaceutical Use by US Astronauts on Space Shuttle Missions."

41. Michael Barratt, "Medical Support for the International Space Station," *Aviation, Space, and Environmental Medicine* 70 (1999): 155–61.

42. Mark R. Campbell and Roger D. Billica, "A Review of Microgravity Surgical Investigations," *Aviation, Space, and Environmental Medicine* 63 (1992): 523–24.

43. Ibid.

44. Daniel S. Goldin, "Steps to Mars," in *Strategies for Mars: A Guide to Human Exploration,* ed. Carol R. Stoker and Carter Emmart (San Diego, Calif.: American Astronautical Society/Univelt, 1996), pp. xi–xix.

45. Robert M. Beattie Jr., "Death in Space," in *The Case for Mars II,* ed. Christopher P. McKay (San Diego, Calif.: American Astronautical Society/Univelt, 1989), pp. 681–93.

4. Life Support

1. A. W. Sloan, *Man in Extreme Environments* (Springfield, Ill.: Charles C. Thomas, 1979).

2. Peter Bond, *Reaching for the Stars: The Illustrated History of Manned Spaceflight* (London: Cassell Publishers, 1996).

3. This discussion of spacecraft is based on Anthony R. Curtis, *Space Almanac,* 2d ed. (Houston: Gulf Publishing, 1992); Bond, *Reaching for the Stars;* Philip Clark, *The Soviet Manned Space Program* (New York: Orion Books, 1988); Arnauld E. Nicogossian and John J. Uri, "Vehicles for Human Spaceflight," in *Space Physiology and Medicine,* ed. Arnauld E. Nicogossian, Carolyn Leach Huntoon, and Sam L. Pool, 3d ed. (Philadelphia: Lea and Febiger, 1994), pp. 109–27.

4. Brian Harvey, *The New Russian Space Programme: From Competition to Collaboration* (Chichester, U.K.: Wiley-Praxis, 1996).

5. Dennis R. Jenkins, *Space Shuttle: The History of Developing the National Space Transportation System* (Cape Canaveral: Dennis R. Jenkins, 1997).

6. David M. Harland, *The Space Shuttle: Roles, Missions, and Accomplishments* (Chichester, U.K.: Wiley-Praxis, 1998).

7. Harvey, *The New Russian Space Programme.*

8. H. S. F. Cooper, *A House in Space* (New York: Bantam Books, 1976).

9. Harvey, *The New Russian Space Programme;* David M. Harland, *The Mir Space Station: A Precursor to Space Colonization* (Chichester, U.K.: Wiley-Praxis, 1997).

10. Bryan Burrough, *Dragonfly: NASA and the Crisis aboard Mir* (New York: HarperCollins, 1998).

11. Harland, *The Mir Space Station.*

12. David A. Brown, "International Space Station Faces Further Delays, Controversy," *Launchspace* (July 1997): 30–34; Charles Gunn, "NASA's Interim Control Module a Potential Life-Saver," *Launchspace* (July 1997): 35–38; John Joss, "Status Report: A Conversation with Boeing's Chief Engineer on the ISS," *Launchspace* (July 1997): 38–42.

13. Penelope Boston, "Moving in on Mars: The Hitchhiker's Guide to Life Support," in *Strategies for Mars: A Guide to Human Exploration,* ed. Carol R. Stoker and Carter Emmart (San Diego, Calif.: American Astronautical Society/Univelt, 1996), pp. 327–61.

14. Ben Bova, *Welcome to Moonbase* (New York: Ballantine Books, 1987).

15. Thomas R. Meyer and Christopher P. McKay, "Using the Resources of Mars for Human Settlement," in *Strategies for Mars: A Guide for Human Exploration,* ed. Carol R. Stoker and Carter Emmart (San Diego, Calif.: American Astronautical Society/Univelt, 1996), pp. 393–443.

16. Boston, "Moving in on Mars."

17. James M. Waligora, Michael R. Powell, and Richard L. Sauer, "Spacecraft Life-Support Systems," in *Space Physiology and Medicine,* ed. Arnauld E. Nicogossian, Carolyn Leach Huntoon, and Sam L. Pool, 3d ed. (Philadelphia: Lea and Febiger, 1994), pp. 109–27.

18. Burrough, *Dragonfly.*

19. Waligora, Powell, and Sauer, "Spacecraft Life-Support Systems."

20. Meyer and McKay, "Using the Resources of Mars for Human Settlement."

21. Robert Zubrin and Richard Wagner, *The Case for Mars* (New York: Free Press, 1996).

22. Stephen Baxter, "The Hidden Ocean: Mining Deep Water on the Moon," *Journal of the British Interplanetary Society* 51 (1998): 75–80.

23. Meyer and McKay, "Using the Resources of Mars for Human Settlement."

24. Robert E. Feeney, *Polar Journeys: The Role of Food and Nutrition in Early Exploration* (Washington, D.C.: American Chemical Society; Fairbanks: University of Alaska Press, 1998).

25. Helen W. Lane and Paul C. Rambaut, "Nutrition," in *Space Physiology and Medicine,* ed. Arnauld E. Nicogossian, Carolyn Leach Huntoon, and Sam L. Pool, 3d ed. (Philadelphia: Lea and Febiger, 1994), pp. 305–16.

26. Lillian D. Kozloski, *US Space Gear: Outfitting the Astronaut.* (Washington, D.C.: Smithsonian Institution Press, 1994).

27. Jack Stuster, *Bold Endeavors: Lessons from Polar and Space Exploration* (Annapolis, Md.: Naval Institute Press, 1996).

28. Ibid.

29. William R. Pogue, *How Do You Go to the Bathroom in Space?* (New York: Tor Books, Tom Doherty Associates, 1991).

30. Burrough, *Dragonfly.*

31. Meyer and McKay, "Using the Resources of Mars for Human Settlement."

32. Robert C. Boyd, Patrick S. Thompson, and Benton C. Clark, "Duricrete Composites Construction on Mars," in *The Case for Mars II,* ed. Christopher P. McKay (San Diego, Calif.: American Astronautical Society/Univelt, 1985), pp. 539–50; Bruce A. Mackenzie, "Building Mars Habitats Using Local Materials," in *The Case for Mars II,* ed. Christopher P. McKay (San Diego, Calif.: American Astronautical Society/Univelt, 1985), pp. 575–86.

33. Amos Banin, "Mars Soil: A Sterile Regolith or a Medium for Plant Growth?" in *The Case for Mars II,* ed. Christopher P. McKay (San Diego, Calif.: American Astronautical Society/Univelt, 1985), pp. 559–72.

34. Freeman Dyson, *Imagined Worlds* (Cambridge: Harvard University Press, 1997).

35. Mark Nelson and William F. Dempster, "Living in Space: Results from Biosphere 2's Initial Closure, an Early Testbed for Closed Ecological Systems on Mars," in *Strategies for Mars: A Guide to Human Exploration,* ed. Carol R. Stoker and Carter Emmart (San Diego, Calif.: American Astronautical Society/Univelt, 1996), 363–90.

36. Martyn Fogg, *Terraforming: Engineering Planetary Environments* (Warredale, Penn.: Society of Automotive Engineers, 1995).

37. James E. Oberg, *New Earths* (New York: New American Library, 1981); Christopher P. McKay, "Terraforming Mars," *Journal of the British Interplanetary Society* 35 (1982): 427–33; Fogg, *Terraforming.*

38. Marshall Savage, *The Millennial Project: Colonizing the Galaxy in Eight Easy Steps* (Boston: Little, Brown, 1994).

39. Christopher P. McKay and Robert H. Haynes, "Implanting Life on Mars as a Long Term Goal for Mars Exploration," in *The Case for Mars IV: Considerations for Sending Humans,* ed. Thomas R. Meyer (San Diego, Calif.: American Astronautical Society/Univelt, 1997), pp. 209–15.

40. D. MacNiven, "Environmental Ethics and Planetary Engineering," *Journal of the British Interplanetary Society* 48 (1995): 441–44.

41. Owen Gwynne and Christopher P. McKay, "Extracting Water from the Martian Soil Using Microwaves," in *The Case for Mars IV: Considerations for Sending Humans,* ed. Thomas R. Meyer (San Diego, Calif.: American Astronautical Society/Univelt, 1997), pp. 149–53; R. A. Mole, "Terraforming Mars with Four War Surplus Bombs," *Journal of the British Interplanetary Society* 48 (1995): 321.

42. Harvey, *The New Russian Space Programme.*

43. Fogg, *Terraforming,* p. 327.

44. Haym Benroya, "An Engineering Perspective on Terraforming," *Journal of the British Interplanetary Society* 50 (1997): 105–8.

5. HABITABILITY

1. Michael Collins, *Liftoff* (New York: Grove Press, 1988).
2. Albert A. Harrison, Nancy J. Struthers, and Bernard J. Putz, "Individual Differences and Mission Parameters and Determinants of Spaceflight Environment Habitability," in *The Case for Mars III*, ed. Carol Stoker, vol. 2 (San Diego, Calif.: American Astronautical Society/Univelt, 1989), pp. 191–99; Albert A. Harrison, Nancy J. Struthers, and Bernard J. Putz, "Mission Destination, Mission Duration, Gender, and Student Perceptions of Space Habitat Acceptability," *Environment and Behavior* 23 (1991): 221–32.
3. Mary M. Connors, Albert A. Harrison, and Faren R. Akins, *Living Aloft: Human Requirements for Extended-Duration Spaceflight*, NASA SP-483 (National Aeronautics and Space Administration, 1985).
4. Victoria Garshnek and Jeri W. Brown, "Human Capabilities in Space Exploration and Utilization," in *Space Physiology and Medicine*, ed. Arnauld E. Nicogossian, Carolyn Leach Huntoon, and Sam L. Pool, 2d ed. (New York: Lea and Febiger, 1989), p. 375.
5. Space Science Board, National Research Council, *Polar Biomedical Research: An Assessment* (Washington, D.C.: National Academy Press, 1987), p. 36.
6. Peter Suedfeld, "Extreme and Unusual Environments," in *Handbook of Environmental Psychology*, ed. Daniel Stokols and Irwin Altman, vol. 1 (New York: Wiley, 1987), pp. 863–87; Peter Suedfeld, "Groups in Isolation and Confinement: Environments and Experiences," in *From Antarctica to Outer Space: Life in Isolation and Confinement*, ed. Albert A. Harrison, Yvonne A. Clearwater, and Christopher P. McKay (New York: Springer-Verlag, 1991), pp. 134–46.
7. Harrison, Struthers, and Putz, "Individual Differences and Mission Parameters and Determinants of Spaceflight Environment Habitability"; Harrison, Struthers, and Putz, "Mission Destination, Mission Duration, Gender, and Student Perceptions of Space Habitat Acceptability."
8. Clayton P. Alderfer, *Existence, Relatedness, and Growth* (New York: Free Press, 1972).
9. Albert A. Harrison and Mary M. Connors, "Human Factors in Spacecraft Design," *Journal of Spacecraft and Rockets* 27 (1990): 478–81.
10. Wolfgang E. Preiser, "Environmental Design Cybernetics: A Relativistic Conceptual Framework for the Design of Space Stations and Settlements," in *From Antarctica to Outer Space: Life in Isolation and Confinement*, ed. Albert A. Harrison, Yvonne A. Clearwater, and Christopher P. McKay (New York: Springer-Verlag, 1991), pp. 147–60.
11. National Aeronautics and Space Administration, *Man-Systems Integration Standards STD-3000*, NASA (March 1987).
12. Molly Elrod, "Considerations of a Habitat Design," *Journal of the British Interplanetary Society* 48 (1995): 39–42.
13. Irwin Altman, *The Environment and Social Behavior* (Monterey, Calif.: Brooks-Cole, 1975); Robert Sommer, *Personal Space* (Englewood Cliffs, N.J.: Prentice-Hall, 1969).

14. Douglas Raybeck, "Proxemics and Privacy: Managing the Problems of Life in Confined Environments," in *From Antarctica to Outer Space: Life in Isolation and Confinement,* ed. Albert A. Harrison, Yvonne A. Clearwater, and Christopher P. McKay (New York, Springer-Verlag, 1991), pp. 317–31.

15. Joshua E. Summit, Susan C. Westfall, Robert Sommer, and Albert A. Harrison, "Weightlessness and Interaction Distance: A Simulation of Interpersonal Contact in Outer Space," *Environment and Behavior* 24 (1992): 617–33.

16. Bryan Burrough, *Dragonfly: NASA and the Crisis aboard Mir* (New York: HarperCollins, 1998).

17. Michael Roberts, "The Use of Inflatable Habitation on the Moon and Mars," in *The Case for Mars III: Strategies for Exploration—General Interest and Overview,* ed. Carol Stoker (San Diego, Calif.: American Astronautical Society/Univelt, 1989), pp. 587–94.

18. Marc M. Cohen, "First Mars Outpost Habitation Strategy," in *Strategies for Mars: A Guide for Human Exploration,* ed. Carol R. Stoker and Carter Emmart (San Diego, Calif.: American Astronautical Society/Univelt, 1996), pp. 465–512.

19. Willy Z. Sadeh and Marvin E. Criswell, "Inflatable Structures for a Lunar Base," *Journal of the British Interplanetary Society* 48 (1995): 33–38.

20. Gary S. Brierly, D. Bryan Neely, and Mark T. Newkirk, "A Remotely Deployable Martian Habitat," in *The Case for Mars IV: Considerations for Sending Humans,* ed. Thomas R. Meyer (San Diego, Calif.: American Astronautical Society/Univelt, 1997), pp. 295–300.

21. Connors, Harrison, and Akins, *Living Aloft;* Albert A. Harrison, Robert Sommer, Nancy Struthers, and Kathleen Hoyt, *Implications of Privacy Needs and Interpersonal Distancing Mechanisms for Space Station Design,* NASA Contractor Report 177500 (August 1988).

22. Albert A. Harrison, Barrett Caldwell, Nancy J. Struthers, and Yvonne A. Clearwater, *Incorporation of Privacy Elements in Space Station Design,* Final Report, NASA NAG 2–431 (May 1988).

23. William R. Pogue, *How Do You Go to the Bathroom in Space?* (New York: Tor Books, Tom Doherty Associates, 1991).

24. Raybeck, "Proxemics and Privacy."

25. Charles A. Berry, "A View of Human Problems to Be Addressed for Long Duration Space Flight," *Aerospace Medicine* 44 (1973): 1136–46.

26. Yvonne A. Clearwater, "Space Station Habitability Research," *Acta Astronautica* 17 (1988): 217–22; Yvonne A. Clearwater and Richard G. Coss, "Functional Esthetics to Enhance Well-Being in Isolated and Confined Settings," in *From Antarctica to Outer Space: Life in Isolation and Confinement,* ed. Albert A. Harrison, Yvonne A. Clearwater, and Christopher P. McKay (New York, Springer-Verlag, 1991), pp. 331–58.

27. Christopher Barbour and Richard G. Coss, "Differential Color Brightness as a Body Orientation Cue," *Human Factors* 30 (1988): 713–17.

28. Raybeck, "Proxemics and Privacy"; National Aeronautics and Space Administration, *Man-Systems Integration Standards STD-3000;* D. Nixon, *Space Station Group Activities Habitability Module Study,* NASA Contractor Report 4010 (November 1986).

29. Clearwater and Coss, "Functional Esthetics to Enhance Well-Being in Isolated and Confined Settings."

30. Benjamin B. Weybrew, "Three Decades of Nuclear Submarine Research: Implications for Space and Antarctic Research," in *From Antarctica to Outer Space: Life in Isolation and Confinement,* ed. Albert A. Harrison, Yvonne A. Clearwater, and Christopher P. McKay (New York: Springer-Verlag, 1991), pp. 103–14.

31. Paul N. Klaus, "Decreasing Stress through the Introduction of Micro-environments," in *From Antarctica to Outer Space: Life in Isolation and Confinement,* ed. Albert A. Harrison, Yvonne A. Clearwater, and Christopher P. McKay (New York: Springer-Verlag, 1991), pp. 359–62.

32. Richard F. Haines, "Windows in Confining Environments," in *From Antarctica to Outer Space: Life in Isolation and Confinement,* ed. Albert A. Harrison, Yvonne A. Clearwater, and Christopher P. McKay (New York: Springer-Verlag, 1991), pp. 349–58.

33. Weybrew, "Three Decades of Nuclear Submarine Research."

34. Clearwater and Coss, "Functional Esthetics to Enhance Well-Being in Isolated and Confined Settings."

35. National Aeronautics and Space Administration, *Man-Systems Integration Standards STD-3000.*

36. T. S. Clark and E. N. Corlett, *The Ergonomics of Workspaces and Machines: A Design Manual* (London: Taylor and Francis, 1984).

37. J. Boud, *Lighting Designs in Buildings* (Senvege, U.K.: Peter Peregrinus, 1973).

38. National Aeronautics and Space Administration, *Man-Systems Integration Standards STD-3000.*

39. Pogue, *How Do You Go to the Bathroom in Space?*

40. This discussion is based on Albert A. Harrison, Barrett Caldwell, Nancy J. Struthers, and Yvonne A. Clearwater, *Incorporation of Privacy Elements in Space Station Design,* NASA Final Report, NAG 2-341 (May 20, 1988).

41. Pogue, *How Do You Go to the Bathroom in Space?*

42. K. F. Willshire and K. D. Leatherwood, *Astronaut Survey of Shuttle Vibroacoustic Environment* PIR No. SD-6 (Hampton, Va.: Acoustics Division, Structures Directorate, Langley Research Center, 1985).

43. National Research Council, Committee on Hearing, Bioacoustics, and Biomechanics, Commission on Behavioral and Social Sciences and Education, *Guidelines for Noise and Vibration Levels for the Space Station* (Washington, D.C.: National Research Council, 1997).

44. Preiser, "Environmental Design Cybernetics."

6. SELECTION AND TRAINING

1. Boeing Defense and Space Group, "The Space Station: It's about Life on Earth" (n.p., n.d.).

2. Paul W. Caro, "Flight Training and Simulation," in *Human Factors in*

Aviation, ed. Earl L. Wiener and David C. Nagel (San Diego, Calif.: Academic Press, 1988), pp. 229–63.

3. Louis F. Fogg and Robert M. Rose, "Use of Personal Characteristics in the Selection of Astronauts," *Aviation, Space, and Environmental Medicine* 66 (1995): 199–205.

4. "NASA Selects 25 Astronaut Candidates: Do Any Live in Your Home Town?" *Final Frontier* (October 1998): p. 14.

5. Http://shuttle.nasa.gov/sts77/factshts/astroqualify.html/.

6. Tom Wolfe, *The Right Stuff* (New York: Farrar, Straus, Giroux, 1969).

7. Http://shuttle.nasa.gov/sts77/factshts/astroqualify.html/.

8. Brian Harvey, *The New Russian Space Programme: From Competition to Collaboration* (Chichester, U.K.: Wiley-Praxis, 1996), p. 40.

9. Marshall Savage, *The Millennial Project* (Boston: Little, Brown, 1994), p. 206.

10. Robert J. Biersner, "Psychological Evaluation and Selection of Divers," in *Physician's Guide to Diving Medicine,* ed. C. W. Schilling and C. B. Carlson (New York: Plenum, 1984).

11. Patricia Santy, *Choosing the Right Stuff: The Psychological Selection of Astronauts and Cosmonauts* (Westport, Conn.: Praeger, 1994).

12. O. G. Edholm and E. K. E. Gunderson, eds., *Polar Human Biology* (London: Wm. Heinnemann, 1973).

13. A. J. W. Taylor, *Antarctic Psychology,* DSIR Bulletin No. 24 (Wellington, N.Z.: Science Information Publishing Center, 1987).

14. Patricia A. Santy and David R. Jones, "An Overview of International Issues in Astronaut Psychological Selection," *Aviation, Space, and Environmental Medicine* 65 (1994): 900–903; Chiaru Sekiguchi, Sei Umikura, Keichi Sone, and Minoru Kume, "Psychological Evaluation of Japanese Astronaut Applicants," *Aviation, Space, and Environmental Medicine* 65 (1994): 920–24.

15. Sheryl L. Bishop, Dean Faulk, and Patricia A. Santy, "The Use of IQ Assessment in Astronaut Screening and Evaluation," *Aviation, Space, and Environmental Medicine* 67 (1996): 1130–37.

16. Brian Harvey, *The New Russian Space Programme: From Competition to Collaboration* (Chichester, U.K.: Wiley-Praxis, 1996), p. 6.

17. Ibid., p. 38.

18. Santy, *Choosing the Right Stuff.*

19. Bryan Burrough, *Dragonfly: NASA and the Crisis aboard Mir* (New York: HarperCollins, 1998).

20. Terry L. McFadden, Robert L. Helmreich, Robert M. Rose, and Louis F. Fogg, "Predicting Astronaut Effectiveness: A Multivariate Approach," *Aviation, Space, and Environmental Medicine* 65 (1994): 904–9; Robert M. Rose, Robert L. Helmreich, Louis Fogg, and Terry J. McFadden, "Assessments of Astronaut Effectiveness," *Aviation, Space, and Environmental Medicine* 64 (1993): 789–94.

21. Mary M. Connors, Albert A. Harrison, and Faren R. Akins, *Living Aloft: Human Requirements for Extended Spaceflight,* NASA SP-483 (National Aeronautics and Space Administration, 1985).

22. H. S. F. Cooper Jr., *Before Liftoff: The Making of a Shuttle Crew* (Baltimore: Johns Hopkins University Press, 1987); http://spaceflight.nasa.gov/shuttle/reference/factsheets/asseltr.html/.

23. Burrough, *Dragonfly*.

24. Cooper, *Before Liftoff*.

25. Burrough, *Dragonfly*.

26. John Young quoted in Cooper, *Before Liftoff*.

27. Cooper, *Before Liftoff*.

28. John Schuessler, personal communication, July 1999.

29. Don Scott, personal communication, July 1999.

30. William A. Hodges, "The Division of Labor and Interstellar Migration," in *Interstellar Migration and the Human Experience*, ed. Ben R. Finney and Eric M. Jones (Berkeley and Los Angeles: University of California Press, 1984), pp. 120–33.

31. J. F. Kubis and E. L. McLaughlin, "Psychological Aspects of Spaceflight," *Transactions of the New York Academy of Science* 30 (December 1967).

7. Stress and Coping

1. "July 16 in History: Between Heaven and Earth," *CNI News*, vol. 5, no. 10, pt. 2 (July 16, 1999), p. 4.

2. "Neil Armstrong Says He Had No Script for First Words on Moon," *Sacramento Bee*, July 17, 1999, p. A12.

3. William Pogue, *How Do You Go to the Bathroom in Space?* (New York: Tor Books, Tom Doherty Associates, 1991).

4. Bryan Burrough, *Dragonfly: NASA and the Crisis aboard Mir* (New York: HarperCollins, 1998).

5. Philip R. Harris, *Living and Working in Space: Human Behavior, Culture, and Organization*, 2d ed. (Chichester, U.K.: Wiley-Praxis, 1996), p. 18.

6. Eugene Cernan and Don Davis, *Astronaut Eugene Cernan and America's Great Race in Space* (New York: St. Martin's Press, 1999).

7. J. R. Godwin, "A Preliminary Investigation into Stress in Australian Antarctic Expeditions," in *Human Factors in Polar Psychology: A Symposium*, ed. A. J. W. Taylor, Polar Symposia 1 (Cambridge: Scott Polar Research Institute, 1991), pp. 9–22.

8. C. D. Spielberger, L. G. Westberry, and G. Greenfield, "A Close Look at Police Stress," *Florida Fraternal Order of Police Journal* (fall 1981): 31–43; T. A. Martelli, L. K. Waters, and J. Martelli, "The Police Stress Survey," *Journal of Police Science and Administration* 17 (1989): 267–73.

9. Desmond Lugg, "Current International Human Factors Research in Antarctica," in *From Antarctica to Outer Space: Life in Isolation and Confinement*, ed. Albert A. Harrison, Yvonne A. Clearwater, and Christopher P. McKay (New York: Springer-Verlag, 1991), pp. 31–42.

10. Benjamin B. Weybrew and Ernest M. Noddin, "Psychiatric Aspects of Adaptation to Long Submarine Missions," *Aviation, Space, and Environmen-*

tal Medicine 50 (1979): 575–80; Benjamin B. Weybrew and Ernest M. Noddin, "The Mental Health of Nuclear Submariners in the United States Navy," *Military Medicine* 44 (1979): 188–91.

11. Frank White, *The Overview Effect* (Boston: Houghton Mifflin, 1987).

12. Donna C. Oliver, "Psychological Effects of Isolation and Confinement of a Winter-Over Group at McMurdo Station, Antarctica," in *From Antarctica to Outer Space: Life in Isolation and Confinement,* ed. Albert A. Harrison, Yvonne A. Clearwater, and Christopher P. McKay (New York: Springer-Verlag, 1991), pp. 224–25.

13. Lawrence A. Palinkas, Peter Suedfeld, and G. Daniel Steel, "Psychological Functioning among Members of a Small Polar Expedition," *Aviation, Space, and Environmental Medicine* 66 (1995): 943–50.

14. Vyacheslav I. Myasnikov and Iltuzar S. Zamaletdinov, "Psychological States and Group Interactions of Crew Members in Flight," in *Humans in Spaceflight,* ed. Carolyn S. Leach Huntoon, Vsevolod V. Antipov, and Anatoliy I. Grigoriev, vol. 3, bk. 2 (Reston, Va.: American Institute of Aeronautics and Astronautics, 1996), pp. 419–31.

15. Peter Suedfeld, "Groups in Isolation and Confinement: Environments and Experiences," in *From Antarctica to Outer Space: Life in Isolation and Confinement,* ed. Albert A. Harrison, Yvonne A. Clearwater, and Christopher P. McKay (New York: Springer-Verlag, 1991), pp. 135–46.

16. Michael Freeman, "Pilot to Try 2nd Record Flight," *Sacramento Bee,* December 29, 1997, p. A4.

17. Lawrence A. Palinkas, "Long-Term Effects of Environment on Health and Performance of Antarctic Winter-Over Personnel," Report 86–3 (San Diego, Calif.: Naval Health Research Center, 1986); Lawrence A. Palinkas, "Group Adaptation and Individual Adjustment in Antarctica: A Summary of Recent Research," in *From Antarctica to Outer Space: Life in Isolation and Confinement,* ed. Albert A. Harrison, Yvonne A. Clearwater, and Christopher P. McKay (New York: Springer-Verlag, 1991), pp. 239–52.

18. Sam L. Pool, Arnauld E. Nicogossian, Edward C. Moselet, John J. Uri, and Larry J. Pepper, "Medical Evaluations for Astronaut Selection and Longitudinal Studies," in *Space Physiology and Medicine,* ed. Arnauld E. Nicogossian, Carolyn Leach Huntoon, and Sam L. Pool, 3d ed. (Philadelphia: Lea and Febiger, 1994), pp. 375–93.

19. C. S. Mullin, H. Connery, and F. Wouters, "A Psychological-Psychiatric Study of an IGY Station in Antarctica (Project Report), United States Navy, Neuropsychiatric Division of the Bureau of Medicine and Surgery," quoted in Arreed F. Barabasz, "Effects of Isolation on States of Consciousness," in *From Antarctica to Outer Space: Life in Isolation and Confinement,* ed. Albert A. Harrison, Yvonne A. Clearwater, and Christopher P. McKay (New York: Springer-Verlag, 1991), pp. 201–9.

20. A. F. Barabasz, "Effects of Isolation on States of Consciousness," pp. 201–9.

21. Marianne Barabasz, "Imaginative Involvement in Antarctica: Applications to Life in Space," in *From Antarctica to Outer Space: Life in Isolation*

and Confinement, ed. Albert A. Harrison, Yvonne A. Clearwater, and Christopher P. McKay (New York: Springer-Verlag, 1991), pp. 209–15.

22. Albert A. Harrison and Joshua Summit, "How 'Third Force' Psychology Might View Humans in Space," *Space Power* 10, no. 2 (1991): 185–203.

23. Anthony Storr, *Solitude: A Return to the Self* (New York: Free Press, 1988).

24. Jerald A. Greenberg and Robert A. Baron, *Behavior in Organizations,* 6th ed. (Upper Saddle River, N.J.: Prentice Hall, 1997), p. 237.

25. Donald K. Slayton and Michael Cassutt, *Deke!* (New York: Forge Books, 1994).

26. Burrough, *Dragonfly.*

27. Pogue, *How Do You Go to the Bathroom in Space?*

28. Brian Harvey, *The New Russian Space Programme: From Competition to Collaboration* (Chichester, U.K.: Wiley-Praxis, 1996), p. 273.

29. E. K. Eric Gunderson and Lawrence A. Palinkas, "Psychological Studies in the US Antarctic Program: An Overview," in *Human Factors in Polar Psychology: A Symposium,* ed. A. J. W. Taylor, Polar Symposia 1 (Cambridge, U.K.: Scott Polar Research Institute, 1991), pp. 5–8.

30. Lawrence A. Palinkas, "Going to Extremes: The Cultural Context of Stress, Illness, and Coping in Antarctica," *Social Science and Medicine* 35 (1992): 651–64.

31. Lawrence A. Palinkas, M. Cravalho, and Deirdre Browner, "Seasonal Variation of Depressive Symptoms in Antarctica," *Acta Psychiatrica Scandinavica* 91 (1995): 423–29.

32. Harvey, *The New Russian Space Programme;* Burrough, *Dragonfly.*

33. Nick Kanas, "Psychological, Psychiatric, and Interpersonal Aspects of Long-Duration Space Missions," *Journal of Spacecraft and Rockets* 27 (1990): 457–63; Patricia A. Santy, "Psychiatric Components of a Health Maintenance Facility (HMF) on Space Station," *Aviation, Space, and Environmental Medicine* 58 (1987): 1219–24; Patricia A. Santy, "Psychological Health Maintenance on Space Station Freedom," *Journal of Spacecraft and Rockets* 27 (1990): 482–85.

34. J. H. Rohrer, "Interpersonal Relations in Isolated Small Groups," in *Psychophysiological Aspects of Spaceflight,* ed. B. E. Flaherty (New York: Columbia University Press, 1961).

35. Gary Daniel Steel and Peter Suedfeld, "Temporal Patterns of Affect in an Isolated Group," *Environment and Behavior* (1991): 749–65; Mary M. Connors, Albert A. Harrison, and Faren R. Akins, *Living Aloft: Human Requirements for Extended Spaceflight,* NASA SP-483 (National Aeronautics and Space Administration, 1985).

36. Robert B. Bechtel and Amy Berning, "The Third Quarter Phenomenon: Do People Experience Discomfort after Stress Has Passed?" in *From Antarctica to Outer Space: Life in Isolation and Confinement,* ed. Albert A. Harrison, Yvonne A. Clearwater, and Christopher P. McKay (New York: Springer-Verlag, 1991), pp. 261–65.

37. Myasnikov and Zamaletdinov, "Psychological States and Group Interactions of Crew Members in Flight."

38. Nick Kanas, "Psychosocial Support for Cosmonauts," *Aviation, Space, and Environmental Medicine* 62 (1991): 353–55; Santy, "Psychiatric Components of a Health Maintenance Facility (HMF) on Space Station"; Santy, "Psychological Health Maintenance on Space Station Freedom."

39. Richard S. Lazarus, *Emotion and Adaptation* (New York: Oxford University Press, 1991); Richard S. Lazarus and Susan Folkman, *Stress, Appraisal, and Coping* (New York: Springer, 1981); Richard S. Lazarus and Bernice Lazarus, *Making Sense of Our Emotions* (New York: Oxford University Press, 1994).

40. Hugo O. Leimann Patt, "The Right and Wrong Stuff in Civil Aviation," *Aviation, Space, and Environmental Medicine* 59 (1988): 955–59.

41. Jeana Yeager and Dick Rutan, with Phil Patton, *Voyager* (New York: Knopf, 1987).

42. Lazarus and Folkman, *Stress, Appraisal, and Coping.*

43. Judith M. Orasanu and Patricia Backer, "Stress and Military Performance," in *Performance under Stress,* ed. J. Driskell and E. Salas (Mahweh, N.J.: Lawrence Erlbaum Associates, 1995), pp. 89–125.

44. R. A. Wertkin, "Stress-Inoculation Training: Principles and Applications," *Journal of Contemporary Social Work* 66 (1985): 611–16.

45. Orasanu and Backer, "Stress and Military Performance."

46. Ibid.

47. Sidney M. Blair, "The Antarctic Experience," in *From Antarctica to Outer Space: Life in Isolation and Confinement,* ed. Albert A. Harrison, Yvonne A. Clearwater, and Christopher P. McKay (New York: Springer-Verlag, 1991), pp. 57–64.

48. Lugg, "Current International Human Factors Research in Antarctica."

49. Blair, "The Antarctic Experience."

50. Kanas, "Psychosocial Support for Cosmonauts"; Myasnikov and Zamaletdinov "Psychological States and Group Interactions of Crew Members in Flight."

51. Santy, "Psychiatric Components of a Health Maintenance Facility (HMF) on Space Station."

52. Gerard Eagan, *The Skilled Helper* (Monterey, Calif.: Brooks-Cole, 1978).

53. Peter Suedfeld, "Stimulus Restriction," in *Human Factors in Polar Psychology, with Some Implications for Space,* ed. A. J. W. Taylor, Polar Symposia 1 (Cambridge, U.K.: Scott Polar Research Institute, 1991), pp. 35–38.

54. Ibid.

55. Patricia A. Santy, *Choosing the Right Stuff: The Psychological Selection of Astronauts* (Westport, Conn.: Praeger, 1994).

56. Donald K. Slayton and Michael Cassutt, *Deke!* (New York: Forge Books, 1994).

57. Howard E. McCurdy, *Space and the American Imagination* (Washington, D.C.: Smithsonian Institution Press, 1997), p. 95.

8. Group Dynamics

1. Eugene Cernan and Don Davis, *Astronaut Eugene Cernan and America's Race in Space: The Last Astronaut on the Moon* (New York: St. Martin's Press, 1999), p. 253.

2. Seward Smith, "Studies of Groups in Confinement," in *Sensory Deprivation: Fifteen Years of Research,* ed. J. P. Zubeck (New York: Appleton-Century-Crofts, 1969), pp. 374–403.

3. R. E. Doll and E. K. E. Gunderson, "Group Size, Occupational Status, and Psychological Symptomatology in an Extreme Environment," *Journal of Clinical Psychology* 27 (1971): 196–98; J. E. Nardini, R. S. Hermann, and J. E. Rasmussen, "Navy Psychiatric Assessment Program in the Antarctic," *American Journal of Psychiatry* 119 (1962): 97–105.

4. Nick A. Kanas and William E. Fedderson, "Behavioral, Psychiatric, and Sociological Problems of Long Duration Missions," NASA TM X-58067 (1971).

5. Roland Radloff and Robert Helmreich, *Groups under Stress: Psychological Research in Sealab II* (New York: Appleton-Century-Crofts, 1968).

6. David M. Harland, *The Mir Space Station: A Precursor to Space Colonization* (Chichester, U.K.: Wiley-Praxis, 1997).

7. Scott Montgomery and Timothy R. Gaffney, *Back in Orbit: John Glenn's Return to Space* (Atlanta: Longstreet, 1999).

8. Howard E. McCurdy, *Space and the American Imagination* (Washington D.C.: Smithsonian Institution Press, 1997).

9. Cynthia Griffin, "Rocket Gals," *Final Frontier* 16 (March-April 1996).

10. Donald K. Slayton and Michael Cassutt, *Deke!* (New York: Forge Books, 1994).

11. Mary M. Connors, Albert A. Harrison, and Faren R. Akins, *Living Aloft: Human Requirements for Extended Spaceflight,* NASA SP-483 (National Aeronautics and Space Administration, 1985).

12. T. A. Heppenheimer, *Colonies in Space* (Harrisburg, Penn.: Stackpole Books, 1977), p. 157.

13. Alan D. Kelly and Nick Kanas, "Crewmember Communication in Space: A Survey of Astronauts and Cosmonauts," *Aviation, Space, and Environmental Medicine* 63 (1992): 721–26; Alan D. Kelly and Nick Kanas, "Communication between Space Crews and Ground Personnel: A Survey of Astronauts and Cosmonauts," *Aviation, Space, and Environmental Medicine* 64 (1993): 795–800.

14. W. Peeters and S. Sciacovelli, "Communication Related Aspects in Multinational Missions: Euromir 94," *Journal of the British Interplanetary Society* 49 (1996): 113–20.

15. Bryan Burrough, *Dragonfly: NASA and the Crisis aboard Mir* (New York: HarperCollins, 1998).

16. David M. Harland, personal communication, January 22, 1999.

17. Harland, *The Mir Space Station.*

18. Ibid.

19. Patricia A. Santy, Albert W. Holland, Laurie Looper, and Regina Mar-

condes-North, "Multicultural Factors in an International Crew Debrief," *Aviation, Space, and Environmental Medicine* 64 (1993): 196–200.

20. Slayton and Cassutt, *Deke!*

21. Jon Krakauer, *Into Thin Air: A Personal Account of the Mount Everest Disaster* (New York: Villard Books, 1997).

22. Jack Stuster, *Bold Endeavors: Lessons from Polar and Space Exploration* (Annapolis, Md.: Naval Institute Press, 1996), pp. 95–115.

23. J. F. Kubis, "Isolation, Confinement, and Group Dynamics in Long Duration Space Flight," *Acta Astronautica* 17 (1972): 55.

24. John M. Nicholas and Larry W. Penwell, "A Proposed Profile of the Effective Leader in Human Spaceflight, Based on Findings from Analog Environments," *Aviation, Space, and Environmental Medicine* 66 (1995): 63–72.

25. H. Clayton Foushee and Robert L. Helmreich, "Group Interaction and Flight Crew Performance," in *Human Factors in Aviation,* ed. Earl L. Wiener and David C. Nagel (New York: Academic Press, 1988), pp. 189–228; John M. Nicholas and H. Clayton Foushee, "Organization, Selection, and Training of Crews for Extended Spaceflight: Findings from Analogs and Implications," *Journal of Spacecraft and Rockets* 27 (1990): 451–56.

26. Robert L. Helmreich, H. Clayton Foushee, R. Benson, and W. Russini, "Cockpit Resource Management: Exploring the Attitude-Performance Linkage," *Aviation, Space, and Environmental Medicine* 57 (1986): 1198–200.

27. Paul Hersey and Kenneth L. Blanchard, *Management of Organizational Behavior: Utilizing Human Resources,* 4th ed. (Englewood Cliffs, N.J.: Prentice-Hall, 1982).

28. Foushee and Helmreich, "Group Interaction and Flight Crew Performance," pp. 189–228; Arleigh C. Merritt and Robert L. Helmreich, "Human Factors on the Flight Deck: The Influence of National Culture," *Journal of Cross-Cultural Psychology* 27 (1996): 5–24.

29. Mary M. Connors, "Communications Issues of Spaceflight," in *From Antarctica to Outer Space,* ed. Albert A. Harrison, Yvonne A. Clearwater, and Christopher P. McKay (New York: Springer-Verlag, 1991), pp. 267–79.

30. Kelly and Kanas, "Crewmember Communication in Space."

31. D. C. Feldman, "The Development of Group Norms," *Academy of Management Journal* 9 (1984): 47–53.

32. Burrough, *Dragonfly.*

33. E. A. Haggard, "Isolation and Personality," in *Personality Change,* ed. Philip Worchel and Donn Byrne (New York: Wiley, 1964), pp. 433–69.

34. Elliot Aronson and Judson Mills, "The Effect of Severity of Initiation on Liking for a Group," *Journal of Abnormal and Social Psychology* 59 (1959): 177–81.

35. V. C. Vroom and P. Yetton, *Leadership and Decision Making* (Pittsburgh: University of Pittsburgh Press, 1973).

36. John Schuessler, personal communication, June 24, 1999.

37. David M. Harland, personal communication, January 1999.

38. Chester M. Pierce, "Theoretical Approaches to Adaptation to Antarctica and Space," in *From Antarctica to Outer Space: Life in Isolation and*

Confinement, ed. Albert A. Harrison, Yvonne A. Clearwater, and Christopher P. McKay (New York: Springer-Verlag, 1991), pp. 125–34.

39. Gloria R. Leon, "Individual and Group Process Characteristics of Polar Expedition Teams," *Environment and Behavior* 23 (1991): 723–48.

40. Vadim I. Gushin, Vladimir A. Efimov, Tatiana M. Smirnova, Alla G. Vinokhodova, and Nick Kanas, "Subject's Perceptions of the Crew Interaction Dynamics under Prolonged Isolation," *Aviation, Space, and Environmental Medicine* 69 (1998): 556–61.

41. Larry W. Penwell, "Problems of Intergroup Behavior in Spaceflight Operations," *Journal of Spacecraft and Rockets* 27 (1990): 464–70.

42. Ben R. Finney, "Scientists and Seamen," in *From Antarctica to Outer Space: Life in Isolation and Confinement,* ed. Albert A. Harrison, Yvonne A. Clearwater, and Christopher P. McKay (New York: Springer-Verlag, 1991), pp. 89–102.

43. David M. Harland, personal communication, January 1999.

44. Ibid.

45. Penwell, "Problems of Intergroup Behavior in Spaceflight Operations."

46. Vadim I. Gushin, Nina S. Zaprisa, Tatiana B. Kolnitchenko, Vladimir A. Efimov, Tatiana M. Smirnova, Alla G. Vinokhodova, and Nick Kanas, "Content Analysis of the Crew Communication with External Communicants under Prolonged Isolation," *Aviation, Space, and Environmental Medicine* 68 (1998): 1093–98.

47. Connors, Harrison, and Akins, *Living Aloft.*

48. Cernan and Davis, *Astronaut Eugene Cernan and America's Race in Space,* p. 177.

49. Henry S. F. Cooper, *A House in Space* (New York: Bantam Books, 1976).

50. William K. Douglas, "Psychological and Sociological Aspects of Manned Spaceflight," *From Antarctica to Outer Space: Life in Isolation and Confinement,* ed. Albert A. Harrison, Yvonne A. Clearwater, and Christopher P. McKay (New York: Springer-Verlag, 1991), pp. 81–89.

51. Jerry M. Linenger, *Off the Planet* (New York: McGraw-Hill, 2000).

9. AT WORK

1. Jerry M. Linenger, *Off the Planet* (New York: McGraw-Hill, 2000), p. 90.

2. Michael C. Greenisen and Victor Reggie Edgerton, "Human Capabilities in the Space Environment," in *Space Physiology and Medicine,* ed. Arnauld E. Nicogossian, Carolyn Leach Huntoon, and Sam L. Pool, 3d ed. (Philadelphia: Lea and Febiger, 1994), pp. 194–209.

3. David M. Harland, *The Mir Space Station: A Precursor to Space Colonization* (Chichester, U.K.: Wiley-Praxis, 1997).

4. Mary M. Connors, Faren R. Akins, and Albert A. Harrison, *Living Aloft: Human Requirements for Extended Spaceflight,* NASA-SP 483 (Na-

tional Aeronautics and Space Administration, 1985); Greenisen and Edgerton, "Human Capabilities in the Space Environment."

5. R. Curtis Graeber, "Aircrew Fatigue and Circadian Rhythmicity," in *Human Factors in Aviation*, ed. Earl L. Wiener and David C. Nagel (San Diego, Calif.: Academic Press, 1988); Jack Stuster, *Bold Endeavors: Lessons from Polar and Space Exploration* (Annapolis, Md.: Naval Institute Press, 1996), pp. 45–47.

6. Stuster, *Bold Endeavors*.

7. Lillian D. Kozloski, *US Space Gear: Outfitting the Astronaut* (Washington, D.C.: Smithsonian Institution Press, 1994).

8. Frederick W. Taylor, *The Principles of Scientific Management* (New York: Harper, 1929).

9. Mary M. Connors, Albert A. Harrison, and Joshua Summit, "Crew Systems: Integrating Human and Technical Subsystems for the Exploration of Space," *Behavioral Science* 39 (1994): 183–212.

10. John Schuessler, personal communication, June 29, 1999.

11. R. Mike Mullane, *Do Your Ears Pop in Space?* (New York: Wiley, 1997), p. 110.

12. G. Harry Stine, *Living in Space: A Handbook for Work and Exploration Stations Beyond the Earth's Atmosphere* (New York: M. Evans, 1997), pp. 132–33.

13. David M. Harland, *The Mir Space Station: A Precursor to Space Colonization* (Chichester, U.K.: Wiley-Praxis, 1997).

14. Gregory Stock, *Metaman: The Merging of Humans and Machines into a Global Superorganism* (New York: Simon and Schuster, 1993).

15. Mary M. Connors, "Crew System Dynamics: Combining Humans and Automation," SAE Technical Paper 891530 (paper presented at the 19th Intersociety Conference on Environmental Systems, San Diego, July 24–26, 1990); Connors, Harrison, and Summit, "Crew Systems"; Albert A. Harrison and Mary M. Connors, "Crew Systems: Theoretical and Practical Issues in the Fusing of Humans and Technology" (paper presented at the American Group Psychotherapy Association, San Antonio, Texas, January 1991).

16. Connors, Harrison, and Summit, "Crew Systems."

17. Raymond J. Halyard, *The Quest for Water Planets: Interstellar Colonization in the 21st Century* (Show Low, Ariz.: American Eagle Publications, 1996).

18. Ronald D. Jones, "Using Robots to Support and Assist Human Explorers on the Surface of Mars," in *The Case for Mars III: Strategies for Exploration—Technical*, ed. Carol R. Stoker (San Diego, Calif.: American Astronautical Society/Univelt, 1989), pp. 519–26.

19. Halyard, *The Quest for Water Planets*.

20. Greenisen and Edgerton, "Human Capabilities in the Space Environment."

21. William Manchester, *The Glory and the Dream: A Narrative History of the United States. 1932–1972.* (Boston: Little, Brown, 1973), pp. 224–25.

10. MISHAPS

1. James E. Oberg, *Uncovering Soviet Disasters: The Limits to Glasnost* (New York: Random House, 1988).

2. Ibid.

3. Ibid., p. 172.

4. Ibid.

5. Malcolm M. McConnell, *Challenger: A Major Malfunction* (New York: Doubleday, 1987); Diane Vaughan, *The Challenger Launch Decision: Risky Technology, Culture, and Deviance at NASA* (Chicago: University of Chicago Press, 1996).

6. David M. Harland, *The Mir Space Station: A Precursor to Space Colonization* (Chichester, U.K.: Wiley-Praxis, 1997).

7. Charles Perrow, *Normal Accidents* (New York: Basic Books, 1984).

8. Ibid.

9. Charles E. Billings and William D. Reynard, "Human Factors in Aircraft Incidents: Results of a 7-Year Study," *Aviation, Space, and Environmental Medicine* 55 (1984): 960–65.

10. C. O. Miller, "System Safety," in *Human Factors in Aviation,* ed. Earl L. Wiener and David C. Nagel (San Diego, Calif.: Academic Press, 1988), pp. 53–78.

11. Elwyn Edwards, "Introductory Overview," in *Human Factors in Aviation,* ed. Earl L. Wiener and David C. Nagel (San Diego, Calif.: Academic Press, 1988), pp. 3–24.

12. David C. Nagel, "Human Error in Aviation Accidents," in *Human Factors in Aviation,* ed. Earl L. Wiener and David C. Nagel (San Diego, Calif.: Academic Press, 1988), pp. 263–301.

13. Edwards, "Introductory Overview."

14. Barry H. Kantowitz and Patricia A. Casper, "Human Workload in Aviation," in *Human Factors in Aviation,* ed. Earl L. Wiener and David C. Nagel (San Diego: Academic Press), pp. 157–85.

15. Terence J. Lyons and Carl G. Simpson, "The Giant Hand Phenomenon," *Aviation, Space, and Environmental Medicine* 60 (1989): 64–66.

16. Ellen Langer, *Mindfulness* (Reading, Mass.: Addison-Wesley, 1989).

17. Kantowitz and Casper, "Human Workload in Aviation."

18. Sidney Blair, "The Antarctic Experience," in *From Antarctica to Outer Space: Life in Isolation and Confinement,* ed. Albert A. Harrison, Yvonne A. Clearwater, and Christopher P. McKay (New York: Springer-Verlag, 1991), pp. 57–64.

19. Bryan Burrough, *Dragonfly: NASA and the Crisis aboard Mir* (New York: Harper Collins, 1998).

20. H. Clayton Foushee and Robert L. Helmreich, "Group Interaction and Flight Crew Performance," in *Human Factors in Aviation,* ed. Earl L. Wiener and David C. Nagel (San Diego, Calif.: Academic Press, 1988), pp. 189–225.

21. I. L. Janis, *Groupthink: Psychological Studies of Policy Decisions and Disasters,* 2d ed. (Boston: Houghton-Mifflin, 1982).

22. McConnell, *Challenger.*

23. Foushee and Helmreich, "Group Interaction and Flight Crew Performance."

24. Barbara G. Kanki, "Cockpit Crew Research," in *Risk Management: Expanding Horizons in Nuclear Power and Other Industries,* ed. Ronald A. Knief (New York: Hemisphere Publishing, 1991).

25. Ashley C. Merritt and Robert L. Helmreich, "Human Factors on the Flight Deck," *Journal of Cross-Cultural Psychology* 22 (1996): 4–24.

26. Ibid.

27. Vaughan, *The Challenger Launch Decision.*

28. Oberg, *Uncovering Soviet Disasters.*

29. Burrough, *Dragonfly.*

30. Ibid.

31. Perrow, *Normal Accidents.*

32. Donald A. Norman. *The Design of Everyday Things* (New York: Basic Books, 1988).

33. Perrow, *Normal Accidents.*

34. Eugene Kranz, *Failure Is Not an Option* (New York: Simon and Schuster, 2000).

35. Leonard David, "Mars: The Media . . . the Masses . . . and the Message," in *Strategies for Mars: A Guide to Human Exploration,* ed. Carol R. Stoker and Carter Emmart (San Diego, Calif.: American Astronautical Society/Univelt, 1996), p. 37.

11. Off Duty

1. David M. Harland, *The Mir Space Station: A Precursor to Space Colonization* (Chichester, U.K.: Wiley-Praxis, 1997), p. 133.

2. Marshall Savage, *The Millennial Project: Colonizing the Galaxy in Eight Easy Steps* (Boston: Little, Brown, 1994).

3. Anthony R. Curtis, ed., *Space Almanac,* 2d ed. (Houston: Gulf Publishing, 1992), p. 172.

4. Jack Stuster, *Bold Endeavors: Lessons from Polar and Space Explorations* (Annapolis, Md.: Naval Institute Press, 1996).

5. Noel Mostert, *Supership* (New York: Knopf, 1974).

6. Jerry M. Linenger, *Off the Planet* (New York: McGraw-Hill, 2000).

7. Mary M. Connors, Albert A. Harrison, and Faren R. Akins, *Living Aloft: Human Requirements for Extended Spaceflight,* NASA SP-483 (National Aeronautics and Space Administration, 1985).

8. Patricia A. Santy, Heidi Kapanka, Jeffrey R. Davis, and Donald F. Stewart, "Analysis of Sleep on Shuttle Missions," *Aviation, Space, and Environmental Medicine* 59 (1988): 1094–97.

9. James E. Oberg and Alcestis R. Oberg, *Pioneering Space: Living on the Next Frontier* (New York: McGraw-Hill, 1986).

10. Connors, Harrison, and Akins, *Living Aloft.*

11. Stuster, *Bold Endeavors.*

12. Gloria R. Leon, Ruth Kanfer, Richard G. Hoffman, and Lonnie Dupre,

"Interrelationships of Personality and Coping in a Challenging Extreme Situation," *Journal of Research in Personality* 25 (1991): 357–71; Pauline Kahn and Gloria R. Leon, "Group Climate and Individual Functioning in an All-Women Antarctic Expedition Team," *Environment and Behavior* 26 (1994): 669–97.

13. Stuster, *Bold Endeavors*.

14. Mary M. Connors, "Communications Issues of Spaceflight," in *From Antarctica to Outer Space: Life in Isolation and Confinement*, ed. Albert A. Harrison, Yvonne A. Clearwater, and Christopher P. McKay (New York: Springer-Verlag, 1991), pp. 267–80.

15. Desmond J. Lugg, "Current International Human Factors Research in Antarctica," in *From Antarctica to Outer Space: Life in Isolation and Confinement*, ed. Albert A. Harrison, Yvonne A. Clearwater, and Christopher P. McKay (New York: Springer-Verlag, 1991), pp. 31–42.

16. B. J. Bluth, "Staying Sane in Space," *Mechanical Engineering* 104 (1982): 24–29.

17. J. A. Hammes and R. T. Osborn, "Survival Research in Group Isolation Studies," *Journal of Applied Psychology* 19 (1965): 418–21; J. A. Hammes and J. A. Watson, "Behavior Patterns of Groups Experimentally Confined," *Perceptual and Motor Skills* 20 (1965): 1269–72; J. A. Hammes, T. R. Ahearn, and J. F. Keith Jr., "A Chronology of Two Weeks' Fallout Shelter Confinement," *Journal of Clinical Psychology* 21 (1965): 452–56.

18. Alan D. Kelly and Nick Kanas, "Communication between Space Crews and Ground Personnel: A Survey of Astronauts and Cosmonauts," *Aviation, Space, and Environmental Medicine* 64 (1993): 795–800.

19. Connors, Harrison, and Akins, *Living Aloft*; Connors, "Communications Issues of Spaceflight."

20. Philip R. Harris, *Living and Working in Space*, 2d ed. (Chichester, U.K.: Wiley-Praxis, 1996).

21. R. A. Isay, "The Submariners' Wives' Syndrome," *Psychiatric Quarterly* 42 (1968): 647–52; C. A. Pearlman Jr., "Separation Reactions of Married Women," *American Journal of Psychiatry* 126 (1970): 946–50.

22. Deke Slayton and Michael Cassutt, *Deke!* (New York: Forge Books, 1994).

23. Robert L. Helmreich, J. A. Wilhelm, and T. E. Runge, "Psychological Considerations in Future Space Missions," in *The Human Factors in Outer Space Production*, ed. Stephen T. Cheston and David L. Winters, Selected Symposium No. 50 (Washington, D.C.: American Association for the Advancement of Science, 1980), pp. 1–23.

12. Space Tourism

1. David M. Harland, *The Mir Space Station: A Precursor to Space Colonization* (Chichester, U.K.: Wiley-Praxis, 1997), p. 186.

2. Toyohiro Akiyama, "The Pleasure of Spaceflight," *Journal of Space Technology and Science* 9 (1993): 21–23.

3. Jim Lovell, "Today an Unmanned Mars Rover, Tomorrow a Family Vacation on the Final Frontier?" *Final Frontier* (1998): 44.

4. Sven Abitzsch, "Prospects of Space Tourism" (paper presented at the 9th European Aerospace Congress, Berlin, May 15, 1996).

5. P. Collins, R. Stockmans, and M. Maita, "Demand for Space Tourism in America and Japan, and its Implications for Future Space Activities," http://www.spacefuture.com/archive/demand _ for _ space _ tourism _ in _ America _ and _ Japan.shtml.

6. J. G. Pearsall, "Space Hotels," *Journal of the British Interplanetary Society* 50 (1997): 67–80.

7. Http://www.space_tourism_society.org/.

8. Leonard David, "Space Tourism: Escape Velocity Vacationing," *Final Frontier* (1998): 29–34.

9. Collins, Stockmans, and Maita, "Demand for Space Tourism."

10. Lawrence G. Lemke, "Artificial Gravity: Design Implications for Mars Vehicles," in *Strategies for Mars: A Guide for Human Exploration,* ed. Carol R. Stoker and Carter Emmart (San Diego, Calif.: American Astronautical Society/Univelt, 1996), pp. 153–66.

11. David, "Space Tourism."

12. Ibid.

13. Harvey Wichman et al., *Suborbital Civilian Space Flight: Design Issues,* Aerospace Psychology Laboratory Report No. 4 (Claremont, Calif.: Claremont-McKenna College, 1999).

14. D. M. Ashford, "A Development Strategy for Space Tourism," *Journal of the British Interplanetary Society* 50 (1997): 59–66.

15. Kohki Isozaki et al., "Vehicle Design for Space Tourism," *Journal of Space Technology and Science* 10 (1994): 22–34.

16. Ashford, "A Development Strategy for Space Tourism."

17. Pearsall, "Space Hotels."

18. Robert L. Haltermann, "Evolution of the Modern Cruise Trade and Its Application to Space Tourism" (paper presented at the Space Transportation Association [STA]—National Aeronautics and Space Administration [NASA] Cooperative Space Act Agreement Study on Space Tourism, November 30, 1996).

19. Ben Bova, *Welcome to Moonbase* (New York: Ballantine Books, 1987).

20. Lovell, "Today an Unmanned Mars Rover, Tomorrow a Family Vacation on the Final Frontier?"

21. Patrick Collins, "Space Activities, Space Tourism, and Economic Growth" (paper presented at the 2nd International Symposium on Space Tourism, Bremen, Germany, April 21–23, 1999).

13. SPACE SETTLEMENTS

1. David G. Schrunk, Burton L. Sharpe, Bonnie L. Cooper, and Madhu Thangevelu, *The Moon: Resources, Future Development, and Colonization* (Chichester, U.K.: Wiley-Praxis, 1999).

2. Wendell W. Mendell, ed., *Lunar Bases and Space Activities in the 21st*

Century (Houston: Lunar and Planetary Institute, 1985); Michael B. Duke, Wendell W. Mendell, and Barney B. Roberts, "Lunar Base: A Stepping Stone to Mars," in *The Case for Mars II*, ed. Christopher P. McKay (San Diego, Calif.: American Astronautical Association/Univelt, 1985), pp. 207–20.

3. Timothy L. Stroop, "Lunar Bases of the 20th Century: What Might Have Been," *Journal of the British Interplanetary Society* 48 (1995): 3–10.

4. Ben Bova, *Welcome to Moonbase* (New York: Ballantine Books, 1987).

5. Robert Zubrin and Richard Wagner, *The Case for Mars* (New York: Free Press, 1996).

6. Robert M. Zubrin and Christopher P. McKay, "Technological Requirements for Terraforming Mars," *Journal of the British Interplanetary Society* 50 (1997): 83–92.

7. Gerard K. O'Neill, *The High Frontier* (New York: Bantam Books, 1978).

8. T. A. Heppenheimer, *Colonies in Space* (Harrisburg, Penn.: Stackpole Books, 1977).

9. O'Neill, *The High Frontier*.

10. Marsha Freeman, "Krafft Ehricke's Extraterrestrial Imperative," *Space Governance* 2, no. 2 (1995): 20–24; C. M. Hempsell, "History of Space and the Limits to Growth," *Journal of the British Interplanetary Society* 51 (1998): 323–36.

11. Marshall T. Savage, *The Millennial Project: Colonizing the Galaxy in Eight Easy Steps* (Boston: Little, Brown, 1994).

12. Richard Terra, "Islands in the Sky: Human Exploration and Settlement of the Oort Cloud," in *Islands in the Sky*, ed. Stanley Schmidt and Robert Zubrin (New York: Wiley, 1996), pp. 95–116.

13. Ibid.

14. J. B. Calhoun, "Space and the Strategy of Life," in *Environment and Behavior: The Use of Space by Animals and Men*, ed. A. H. Tesser (New York: Plenum, 1971), pp. 329–87.

15. Irwin Altman, *The Environment and Social Behavior* (Monterey, Calif.: Brooks-Cole, 1975).

16. Jim Dator, "Space Settlements and New Forms of Governance," A Space and Society Lecture, Masters in Space Studies, International Space University (n.d.), http://www.soc.hawaii.edu/future/dator/space/spacesettlements.html/.

17. Alvin Rudoff, *Societies in Space* (New York: Peter Lang Publishing, 1996).

18. Donald Scott, "Keeping the Peace in Space: A Neighborhood Model for a Community-Based Conflict-Resolution-Oriented Justice System," in *From Antarctica to Outer Space: Life in Isolation and Confinement*, ed. Albert A. Harrison, Yvonne A. Clearwater, and Christopher P. McKay (New York: Springer-Verlag, 1991), pp. 363–72.

19. F. Kenneth Schwetje, "Justice in the Antarctic, Space, and Military," in *From Antarctica to Outer Space: Life in Isolation and Confinement*, ed. Albert A. Harrison, Yvonne A. Clearwater, and Christopher P. McKay (New York: Springer-Verlag, 1991), pp. 383–94.

20. Dator, "Space Settlements and New Forms of Governance."

21. Zubrin and Wagner, *The Case for Mars.*

22. Howard E. McCurdy, *Space and the American Imagination* (Washington, D.C.: Smithsonian Institution Press, 1997).

23. Magorah Maruyama, "Designing a Space Community," *Futurist* (October 1976): 239.

24. Rudoff, *Societies in Space.*

25. James E. Oberg and Alcestis Oberg, *Pioneering Space: Living on the Next Frontier* (New York: McGraw-Hill, 1986).

26. Stroop, "Lunar Bases of the 20th Century."

27. Dator, "Space Settlements and New Forms of Governance."

14. INTERSTELLAR MIGRATION

1. Eugene F. Mallove and Gregory L. Matloff, *The Starflight Handbook: A Pioneer's Guide to Interstellar Travel* (New York: John Wiley and Sons, 1989).

2. Ibid.

3. Steven Dole, *Habitable Planets* (Santa Monica, Calif.: RAND Corporation, 1964).

4. Martyn Fogg, "A Planet Dweller's Dream," in *Islands in the Sky,* ed. Stanley Schmidt and Robert Zubrin (New York: Wiley, 1996), pp. 143–67.

5. Raymond J. Halyard, *The Quest for Water Planets: Interstellar Space Colonization in the 21st Century* (Show Low, Ariz.: American Eagle Publications, 1996).

6. Richard Brook Cathcart, "Seeing Is Believing: Planetographic Data Display on a Spherical TV," *Journal of the British Interplanetary Society* 50 (1997): 103–4.

7. Mallove and Matloff, *The Starflight Handbook;* John H. Mauldin, *Prospects for Interstellar Travel* (San Diego, Calif.: American Astronautical Society/Univelt, 1992).

8. James A. Dewar, "Atomic Energy: The Rosetta Stone of Space Flight," *Journal of the British Interplanetary Society* 47 (1994): 199–206.

9. Halyard, *The Quest for Water Planets.*

10. Marc G. Millis, "Emerging Possibilities for Space Propulsion Breakthroughs," *Interstellar Propulsion Society Newsletter* 1 (1995): 1; Marc G. Millis, "Breakthrough Propulsion Physics Program: A White Paper" (NASA Lewis Research Center, Cleveland, 1998).

11. Dewar, "Atomic Energy."

12. Marshall Savage, *The Millennial Project: Colonizing the Galaxy in Eight Easy Steps* (Boston: Little, Brown, 1994).

13. Ibid.

14. J. D. Bernal, *The World, the Flesh, and the Devil* (London: Methuen, 1929), cited in Edward Regis Jr., "The Moral Status of Multigenerational Interstellar Exploration," in *Interstellar Migration and the Human Experience,* ed. Ben R. Finney and Eric M. Jones (Berkeley and Los Angeles: University of California Press, 1994), pp. 248–58.

15. Halyard, *The Quest for Water Planets.*

16. A. Hannson, "From Microsystems to Nanosystems," *Journal of the British Interplanetary Society* 51 (1998): 123–26; Al Globus, David Bailey, Jie Han, Richard Jaffe, Creon Levit, Ralph Merkle, and Deepak Srivastaca, "Aerospace Applications of Molecular Nanotechnology," *Journal of the British Interplanetary Society* 51 (1998): 145–52.

17. Mauldin, *Prospects for Interstellar Travel*; Regis, "The Moral Status of Multigenerational Interstellar Exploration"; Halyard, *The Quest for Water Planets.*

18. Regis, "The Moral Status of Multigenerational Interstellar Exploration."

19. Ibid.

20. Millis, "Emerging Possibilities for Space Propulsion Breakthroughs"; Millis, "NASA Breakthrough Propulsion Physics Program."

21. Ibid.

22. B. Haisch, A. Rueda, and H. E. Puthoff, "Advances in the Proposed Electromagnetic Zero-Point Field Theory of Inertia" (paper presented at the 34th AIAA/ASME/SAE/ASEE Joint Propulsion Conference and Exhibit, Cleveland, Ohio, July 13–15, 1998).

23. Millis, "Emerging Possibilities for Space Propulsion Breakthroughs."

24. Ben R. Finney and Eric M. Jones, eds., *Interstellar Migration and the Human Experience* (Berkeley and Los Angeles: University of California Press, 1984).

25. J. B. Birdsell, "Biological Dimensions of Small, Human Founding Populations," *Interstellar Migration and the Human Experience,* ed. Ben R. Finney and Eric M. Jones (Berkeley and Los Angeles: University of California Press, 1984), pp. 110–19.

26. William A. Hodges, "The Division of Labor and Interstellar Migration," in *Interstellar Migration and the Human Experience,* ed. Ben R. Finney and Eric M. Jones (Berkeley and Los Angeles: University of California Press), pp. 120–33.

27. James W. Valentine, "The Origins of Evolutionary Novelty and Galactic Colonization," in *Interstellar Migration and the Human Experience,* ed. Ben R. Finney and Eric M. Jones (Berkeley and Los Angeles: University of California Press), p. 274.

28. Douglas W. Schwartz, "The Colonizing Experience: A Cross-Cultural Perspective," in *Interstellar Migration and the Human Experience,* ed. Ben R. Finney and Eric M. Jones (Berkeley and Los Angeles: University of California Press, 1984), pp. 234–45.

29. Albert A. Harrison, *After Contact: The Human Response to Extraterrestrial Life* (New York, Plenum, 1997).

30. Halyard, *The Quest for Water Planets.*

31. Ray Norris, "Can Science Survive a SETI Detection?" (paper presented at the International Conference on SETI in the 21st Century, UWS Macarthur, Sydney, Australia, 1998).

32. Declan J. O'Donnell, "Metaspace: A Design for Governance in Outer Space," *Space Governance* 1 (January 1994): 8–15.

33. John Mueller, *Retreat from Doomsday: The Obsolescence of Modern War* (New York: Basic Books, 1988).

34. Thomas R. Cusack and Richard A. Stoll, "Collective Security and State Survival in the Interstate System," *International Studies Quarterly* 38 (1994): 33.

35. John Keegan, *A History of Warfare* (New York, Vintage Books, 1994).

36. Michael H. Hart, "Interstellar Migration, The Biological Revolution, and the Future of the Galaxy," in *Interstellar Migration and the Human Experience,* ed. Ben R. Finney and Eric M. Jones (Berkeley and Los Angeles: University of California Press, 1984), pp. 278–91.

37. Haisch, Rueda, and Puthoff, "Advances in the Proposed Electromagnetic Zero-Point Field Theory of Inertia."

38. Frank Tipler, *The Physics of Immortality* (New York: Doubleday, 1994).

39. Michael N. Mautner, "Directed Panspermia. 3. Strategies and Motivation for Seeding Star Forming Clouds," *Journal of the British Interplanetary Society* 50 (1997): 93–102.

15. Restoring the Dream

1. Roger Handberg, "The Art of the Possible: Economic Prospects for Space, Fantasy and Practicality," *Journal of the British Interplanetary Society* 49 (1996): 381–86.

2. Timothy L. Stroop, "Lunar Bases of the 20th Century," *Journal of the British Interplanetary Society* 48 (1995): 3–10.

3. W. D. Kay, *Can Democracies Fly in Space? The Challenge of Revitalizing the US Space Program* (Westport, Conn.: Praeger, 1995).

4. Howard E. McCurdy, *Space and the American Imagination* (Washington, D.C.: Smithsonian Institution Press, 1997).

5. Ibid.

6. Leonard David, "Mars: The Media . . . the Masses . . . and the Message," in *Strategies for Mars: A Guide to Human Exploration,* ed. Carol R. Stoker and Carter Emmart (San Diego, Calif.: American Astronautical Society/ Univelt, 1996), p. 37.

7. McCurdy, *Space and the American Imagination.*

8. Albert A. Harrison and Robert Bell, "Building Support for the Manned Exploration of Mars" (paper presented at "The Case for Mars VI," Boulder, Colorado, July 1996).

9. Handberg, "The Art of the Possible."

10. Michael A. G. Michaud, *Reaching for the High Frontier: The American Pro-Space Movement, 1972–84* (Westport, Conn.: Praeger, 1986).

11. Kay, *Can Democracies Fly in Space?*

12. D. Rubashkin, "Who Killed the Grand Tour? A Case Study in the Politics of Funding Expensive Space Science," *Journal of the British Interplanetary Society* 50 (1997): 177–84.

13. Howard E. McCurdy, *Inside NASA: High Technology and Organiza-*

tional Change in the US Space Program (Baltimore: Johns Hopkins University Press, 1993).

14. Joan Lisa Bromberg, *NASA and the Space Industry* (Baltimore: Johns Hopkins University Press, 1999).

15. Howard E. McCurdy, *The Space Station Decision: Incremental Politics and Technological Choice* (Baltimore: Johns Hopkins University Press, 1990).

16. Philip R. Harris, *Living and Working in Space: Human Behavior, Culture, and Organization*, 2d ed. (Chichester, U.K.: Wiley-Praxis, 1996).

17. On the former, see National Commission on Space, *Pioneering the Space Frontier* (New York: Bantam, 1986); on the latter, see B. Alexander Howerton, *Free Space: Real Alternatives for Reaching Outer Space* (Port Townsend, Wash.: Loompanics Unlimited, 1995).

18. Howerton, *Free Space.*

19. Haym Benroya, "Economically Viable Lunar Development and Settlement," *Journal of the British Interplanetary Society* 30 (1997): 324.

20. Marshall T. Savage, *The Millennial Project: Colonizing the Galaxy in Eight Easy Steps* (Boston: Little, Brown, 1994).

21. Freeman Dyson, *Imagined Worlds* (Cambridge: Harvard University Press, 1997).

22. Kay, *Can Democracies Fly in Space?*

23. Handberg, "The Art of the Possible," p. 382.

24. Howerton, *Free Space.*

25. Declan J. O'Donnell, "Survey of the Top Ten Policy Problems of 1995," *Space Governance* 2 (1995): 40–44; Philip R. Harris, "Legal Space Frontier Challenges," *Space Governance* 4 (1997): 48–51.

26. Alan Wasser, "The Law That Could Make Privately Funded Space Settlement Profitable," *Space Governance* 5 (1998): 55–57.

27. Richard P. McBrien, *Catholicism* (Minneapolis: Winston Press, 1980).

Index

313

Text:	10/13 Sabon
Display:	Scala Sans
Composition:	Binghamton Valley Composition
Printing and binding:	Haddon Craftsmen
Index:	Marcia Carlson